The
Life of a
Boat

The author

Graeme Ewens is a photojournalist, author and editor of magazines and non-fiction books, with a particular interest in maritime affairs (amongst other things). His works have been published and/or syndicated on five continents and in many languages. From 2008-2015 he published *Harwich Ahoy!* for the Harwich Lifeboat, and witnessed the arrival of the *W&S*, which he documented and helped to de-caulk. He has owned several small commercial workboats and now lives in Falmouth.

He is a contributor to *Maritime Journal* and has been published in *Lloyd's List, Ships Monthly, Port of London News, International Tug & Salvage, Anglia Afloat, Thames Guardian, Waterways World, Classic Boat, WSS News, Shipping News Clippings*, as well as national dailies including *The Guardian* and regional papers such as the London *Evening Standard, Eastern Daily Press, Western Morning News, West Briton* and international publications.

Published in collaboration with Elaine Trethowan (Bawden), Press Officer at Penlee Lifeboat, and Captain Rod Shaw MBE, ex-LOM Harwich Lifeboat.

© Graeme Ewens 2021

British Library Cataloguing-in-Publication Data
A catalogue record for this book is available from the British Library

ISBN 978-0-9523655-3-2

Typeset in Bookman Old Style and Gill Sans
Printed by Booths Print, Penryn, Cornwall TR10 8AA

This edition limited to 1,000 copies. A proportion of any proceeds will be shared between the restoration of the *W&S* and Penlee Lifeboat station.

For sales and trade enquiries contact: lifeofaboaton736@gmail.com

The Life of a Boat

Dedicated to the crew of the Penlee Lifeboat with Respect!!

The 'Nautobiography' of a 90-year-old lifesaver

Graeme Ewens

Graeme Ewens

2021

BUKU PRESS

2021

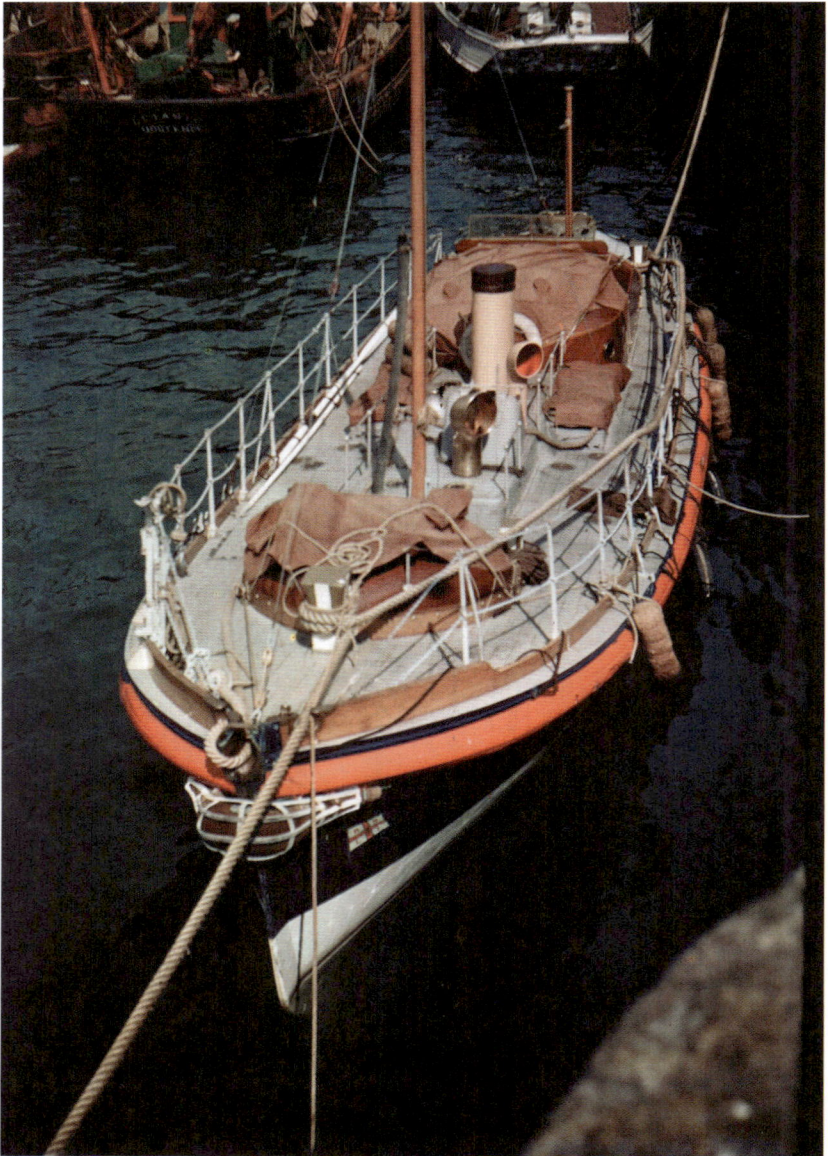

An early colour photograph shows the W&S moored at Newlyn harbour

CONTENTS

Foreword

THE subject of this 'nautobiography', the *W&S*, is one of the longest-serving vessels in the history of the Royal National Lifeboat Institution (RNLI), having been in service at the Penlee station in Mount's Bay, Cornwall from 1931 to 1960, and then kept in the Reserve fleet in northern latitudes until she was sold out of service in 1970. She has since enjoyed semi-retirement as a motor cruiser, spending her 90th birthday undergoing continued restoration in the East of England.

This then is the life story of 'our' Boat, its background and the people who worked her, their extended families, and the vessels she came in contact with — often literally, bumping alongside in all kinds of weather. We believe we have listed every Coxswain who took charge during the boat's career in service with the RNLI, and the subsequent owners and custodians who have taken the helm during the boat's after life. Connections have also been made with people, vessels and places linked to the *W&S* through varying degrees of separation.

Coincidental connections are what brought this author to the current project. I had a long-time connection with Cornwall, and as a photojournalist and editor of illustrated books I had often proposed a book on the 'shipwreck experience' — nothing to do with the glint of sunken treasure or the serenity of wreck diving, but an attempt to get an understanding of the 'adventures and perils' endured by mariners. I have had my share of the former but little experience of such perilous extremes of human experience. My grandfather was shipwrecked twice but without having to endure that catastrophic level of discomfort, this project has provided a window into the lives of lifesavers. As the owner/operator of small commercial workboats, and living on the waterfront in various locations, my contact with lifeboat crews has enabled me to delve into their sub-culture. Like many such self-contained communities it may be 'exclusive' but it is not exclusionary; lifeboat people are open and most welcoming to those who show a genuine interest.

The motivation and the bulk of research undertaken in completing this project has been provided by the Press Officer at the Penlee Lifeboat, Elaine Trethowan (Bawden), whose idea it was to commission the book. Elaine's privileged access to local collections and national archives has enabled her to dig up some invaluable material, while her connection with the Mousehole community could not be more solid: daughter of lifeboat man Nimrod Bawden, she grew up in the harbour village, counting Mechanic Johnny Drew and Cox'n Jack Worth as virtual uncle and grandfather.

The story of the boat's life-saving activities has been compiled from the station's record books and deck log, which Elaine has safeguarded, along with feature articles from the The *Lifeboat* Journal and press reports. The greater part of this history covers the three decades the boat spent in Cornwall, during which the whole pattern of marine technology, industrial development, global trade and international relations changed so fundamentally. The *W&S* was launched in February 1931, eight years before the start of the Second World War, when sailing ships were being superseded by steam-driven iron, and later steel, vessels, and she left the Southwest of England in the first decade of the Jet Age.

Much of the detail in any history is gleaned from dusty archives,

Stalwarts of the Penlee lifeboat and the Mousehole community, Elaine and Nim Bawden

borrowed truths, hindsight and hearsay. Thankfully, a wealth of the Penlee branch's archive has been preserved, and the extensive chapter on the history of the boat's time there is presented in a format that opens most entries with the curt, deadpan delivery of the Returns of Services log books: the initial report, instructions and action taken, sometimes augmented with the more formal rhetoric of official statements. These stories are brought to life, wherever possible, by the more human content: the response, reaction and reflection of crew, witnesses and survivors, plus additional background information to establish the context.

Sadly, some of the central RNLI archives have been destroyed over the years, as have the records of the two local shipyards which carried out maintenance and refits on our boat: Falmouth Boat Company and Mashford's at Cremyll. The archives of our boat's builder, J.S.White, have been dispersed, but thanks to David Williams' book *Maritime Heritage: White's of Cowes* we have most of the yard's history. The Trinity House archives have also been depleted. However, we are fortunate to have the first-hand experience of the longest-serving member of our boat's crew, and the man who would have known her better than any other, the Motor Mechanic Johnny Drew, who spent 29 years on the boat, and in retirement in the 1980s talked to a local oral history group. Drew speaks from a personal, often emotional, viewpoint and we hear his own experiences of some of the *W&S*'s most dramatic actions. And as he says more than once: "The truth can always be re-told. It happened."

As a principal witness Drew has become an inadvertant contributor to our story, and in doing so this modest man has been revealed as a true hero, totally dedicated to the lifeboat service and to his boat. Drew's recollections occupy several early pages of our story and are then interspersed throughout the long section on the boat's services.

Other conversations with coxswains and their descendants have also come to light, and the author was privileged to meet Nim Bawden, mechanic on the ill-fated *Solomon Browne*, who served a shout on the *W&S*, making him the last surviving crew man on our boat. Before he sadly passed away in 2020 Nim gave some more insights into the lifeboat community and he found time to read through the first draft of this book, which gave the author great encouragement and confidence to proceed with the project.

The story of the Penlee lifeboat has been the subject of several books, particularly since the tragic story of the *Solomon Browne*, successor to the *W&S*. Books on the station's history by Mike Sagar-Fenton and Rachael Campey have provided authoritative detail, as have the many volumes published specifically on Cornish wreck and rescue by Grahame Farr, Cyril Noall, Sheila Bird and Richard Larn; and disasters at sea, the most significant being the *Shipwreck Index of the British Isles Vol I* by Richard and Bridget Larn, and published by Lloyd's Register.

Recollections of participants such as Naval officer Richard Perrin and the family of lighthouse keeper Charlie Cherrett add a fresh perspective on well-told tales, while the shipwreck survivor Ralph Richard's experience is extremely poignant. The story of the boat's later years as a leisure cruiser in and out of Northern Ireland was provided by the co-owner at the time, Shirley Bodèll, and the renowned lifeboat expert Quinton Nelson filled in some of its later history. It proved difficult to glean much detail about the boat's time in Falmouth, although Gordon Burns at Flushing and Sam Heard at Mylor confirmed the basic chronology.

The detailed information bringing the story into the 21st Century, is provided by the boat's current owner, Captain Rod Shaw MBE, who acquired ON736 in 2013, and continues with the never-ending task of resurrecting and maintaining the old lady. I had the pleasure of working alongside Rod in Harwich for eight years, during which time we published *Harwich Ahoy!*, a newsletter that grew into a 48-page magazine, which contributed a decent amount to the local station, whose crew I got to know well. Coincidentally, just as Rod was negotiating the purchase of the *W&S*, I was planning to move back to Cornwall, and before relocating I was able to document the first stages of the boat's resurrection at Harwich as the hull was stripped, and I helped in small part with the recaulking. To run my hands over the bare mahogany of that hull was inspirational.

General thanks for providing background information go to Jonathan and Ed Coode for the picture and information on the donor Winifred Alice Coode. Mark Waltham, ex-owner and restorer of the *Laura Moncur* cast his expert eye over the build chapter, and Iain MacQuarrie of MacSalvors contributed details on the salvage of HMS *Warspite*. And we are particularly grateful to the maritime artist Geoffrey Huband, a friend of Johnny Drew, who kindly gave permission to reproduce his paintings of HMS *Anson* at Loe Beach and HMS *Warspite* on the rocks. Similar thanks go to the family of Brian J. Jones for permission to use his painting of the *Britannia* v *Vigilant* race at Mount's Bay.

In helping to trace data and source images and material we are indebted to family and friends of the Penlee lifeboat crew: Eleanor Driscoll at the RNLI; Kathryn Preston, Archivist G.L. Watson & Co.; Ian Smith at Chatham Historic Lifeboat Collection and Gravesend RNLI; Neil Jones at Trinity House; Kevin Camidge of CISMAS; Coxswains and Press officers at Buckie, Aberdeen, Thurso, Broughty Ferry lifeboat stations and Cromarty Archive. Picture credits and digital resources are covered in the appendix. Other publications are included in the bibliography.

Names and Numbers

Before setting out on the long and event-filled history of this memorable boat some identifying markers need to be established. Firstly, several names and numbers referred to in the text should be explained. Although known as the Penlee lifeboat, the station on the north west corner of Mount's Bay has been variously located at Penzance, Newlyn, and Penlee Point, adjacent to the village of Mousehole, the home of the Penlee crew and guardian of the station's proud heritage, indelibly so following the disaster that befell our boat's successor, the *Solomon Browne*. Although still referred to as the Penlee Lifeboat, the current 21st Century boat is housed at the fish dock in Newlyn, some two miles to the north of the old Penlee boathouse. Somewhat confusingly, there is another Penlee Point almost 100 miles to the east of Cornwall, at the mouth of Plymouth Sound, close to the Cremyll shipyard where the *W&S* was often sent for maintenance.

Throughout history the majority of RNLI lifeboats have been named after the generous donors whose bequests have paid for the construction of the boat. In some cases initials have been used rather than the full names of the donor(s). And, by the way, all the Institute's income comes from private or personal sources, with no government or official funding.

The *W&S* was financed by legacies received from Miss Winifred Alice Coode of Launceston, and Miss Ellen Young, of Twickenham, who nominated Captain Sydney Webb. These were both unmarried ladies from distinguished families, wealthy enough to leave substantial legacies. Their contributions were not sufficient to fund a boat each, however, and their money was not used for some 20-odd years, until both sums were combined to finance 'our' lifeboat, which was named using the first initials of the two nominees. The full names of the donors were inscribed on a plate mounted inside the lifeboat, which was removed at some stage during its conversion. The agreed initials, for Winifred and Sydney, were originally written on the boat as *'W and S'* but later photographs (see page 157 bottom right) show it with the ampersand as *W&S*, which we prefer for typographic style and legibility.

It was not uncommon to abbreviate the donors' names; there was a previous boat built in 1929 by the same builders J.S.White named *J&W.* In 1930 another Watson was named *G.W.* and the following year there was *J.H W.* There were at least 19 Watson class boats so named in those decades from the yard of J.S.White alone. Two of the relief boats which stood in for the *W&S* also had short initial names: *B.A.S.P.* served at Penlee in 1940 and *M.O.Y.E.* was on station in 1950. However, there was, coincidentally, another *W&S* vessel working out of Newlyn at the same time as our boat. This was a trawler PZ191, named *W&S* after William and Sara Stevenson, which was itself credited with bringing in two casualty vessels.

The use of initials rather than full names was most frequent during the 1920s and early Thirties. It may seem facetious to suggest the reason to use initials rather than the full names was to save paint, but in those days the lettering was applied in gold leaf and as Britain was suffering economically during the financial depressions of the 1920s, that might have been a good reason to abbreviate the names of donors.

Lifeboat crew, employees and enthusiasts generally refer to the vessels by their Official Number (ON) designated by the RNLI, and our subject was ON736. (The larger Identification Numbers seen on the hulls of current lifeboats were not introduced until much later.) Sticklers for maritime detail also like to know the yard number, given by the builder in sequence as keels are laid down in the construction yard. The 45ft 6in Watson cabin-class boat, to become known as *W&S*, was yard number 1705 in the list of builds from J. Samuel White, at Cowes in the Isle of Wight. That number is branded into the hull's stem. During Rod Shaw's work stripping down the innards of the boat another number came to light: the Commercial Vessel registration number 162590. And since being owned by Rod, the *W&S* has gained yet another identifying number after being placed on the Register of Historic Ships: Certificate no 2996.

The W&S manoeuvring in Newlyn harbour, probably in the 1950s

Winifred and Sydney

Miss Winifred Alice Coode, (1865-1906), had left her legacy of £2,000 for a boat to be built in her name. This would have been equivalent to £3,700 in 1930 when the boat was commissioned — almost exactly half of the total build cost in the RNLI's bought ledger for 1931 (roughly £245,000 in 2020). She is buried at Werrington Parish Church, near Launceston, with the inscription 'In Loving Memory of Winifred Alice Coode of Polapit Tamar, who died on the 8th of August 1906'. Polapit Tamar, the family home, is a large country house built in 1866 by Edward Coode, a local magistrate and High Sheriff of Cornwall for 1868. It was extended in 1901-03 for Edward's son and Winifred's brother, army Captain Richard Carlyon Coode, who inherited the estate on the death of their father in 1894. The house was later turned into a hotel and then a boarding school, before it was converted into apartments.

rin/red A. Coode.

Winifred and her home at Polapit and (below) Vice Admiral Sir John Coode

Coode family members say Winifred (nicknamed Judy) was much loved in the local community, raising money for local causes by organising balls and parties. The magazine *Church Bells* reported in 1898 that her brother Richard, Chairman of Werrington Parish council, campaigned to purchase a complete ring of eight new bells for the parish church of St Martin and St Giles, which should be 'amongst the best in the county. Winifred A. Coode was elected secretary of the executive committee to fund and install the new bells. The Coode family provided six of the bells and all fittings, to the total of £3,291.10s. The weight of the bells is 61 cwt., the tenor being 14 cwt.

PHOTO CHRIS BARNARD

On each bell is inscribed the name of its donor, including one for Winifred Alice.'

Many of her close male relatives were army officers but Judy's extended family also had nautical connections, including two admirals. Her great uncle Vice Admiral Sir John Coode (1779-1858) joined the Navy in 1793 as an AB and moved swiftly through the ranks. By 1814 he was Flag Captain to Rear Admiral Charles Vinicombe Penrose, whose daughter he married. He finally retired as Admiral in charge of victualling at Plymouth Naval Dockyard. His son Trevenen also became a Vice Admiral. Winifred would probably have holidayed with his children, her cousins, at their houses at Lerryn, near the River Fowey.

In 1908 a book of her verse, *The Legend of the Tamar and other poems*, was published posthumously for private circulation. The final poem is titled *My Resting Place*:

High up, upon the western cliff,
Beside the Cornish sea,
Far, far away when I am dead,
There shall you bury me.

No granite cross, no mark of stone,
Above me shall you set;
But let the wild sea-roses run
And wreath their coronet.

Captain Sir (John) Sydney Webb. Miss Ellen Young left her legacy of £650 (£1,170 in 1930) for a boat to be named Sydney, after Captain Sir (John) Sydney Webb, KCMG, JP, DL (1816-98). Miss Young was the sister of Adeline Young, who married Capt Webb in 1849 but died, childless, in 1851. Miss Ellen succeeded Capt Webb as owner of Shirley House in Twickenham after his death in 1898. She died in 1909 and the house was demolished one year later.

Captain (John) Sydney Webb was a naval officer, born in 1814 in London. He was the son of Admiral Charles Webb. Following his Navy career, in 1848 he was elected a Younger Brother of Trinity House, the body overseeing lighthouses around the English coast. In 1857 he became an Elder Brother and in 1883 was elected Deputy-Master, succeeding Admiral Sir Richard Collinson, a position supported by nine active Elder Brethren elected from the ranks of about 300 Younger Brethren. These must be Master Mariners in the Merchant Service or Royal Navy officers of at least Lieutenant-Commander rank. There are a number of honorary members who are not necessarily seamen. The Deputy Master is the active head of Trinity House, while the Mastership is an unpaid position held by a distinguished lay person (currently HRH Princess Anne).

Captain Webb was created Knight Commander of the Order of St Michael and St George in 1889, and when Middlesex County Council was formed he was chosen as an alderman, serving until 1895. He sat for some time on the Thames Conservancy Board as one of the two representatives of Trinity House and he was on the board of the Royal Exchange Assurance Corporation. He was also a member of the Hakluyt Society, established to publish and print 'rare and valuable Voyages, Travels, Naval Expeditions and other geographical records, from an early period to the beginning of the 18th century', who thanked him for helping its members to access the archives of Trinity House.

In 1893 Captain Webb was awarded the Grand Cross of the Order of Saxe Ernestine and, in 1894, he and Elder Brother Captain Vyvyan travelled on the Trinity Steam Yacht *Galatea* to the island of Heligoland in the North Sea to hand over the island and its lighthouse to the German government, after it having been a British possssion under the care of Trinity House since 1807. In exchange the Germans

Following a distinguished Naval career Capt Webb became Deputy Master of Trinity House and made use of TSY Galatea on his inspection tours. The yacht was sold in 1895

relinquished control of Zanzibar to Britain in the carving up of European colonies in the 'scramble for Africa'. Later that year Capt Webb was one of the dignitaries presented to the Royal Party at the opening of London's Tower Bridge.

In 1898 he was taken ill aboard his yacht and died soon after at his home in Twickenham.

The subject of this Nautobiography, the W&S (ON736) takes to the water in February, 1931

Adventures and Perils

**"The sea is a field of tragedy; but it is also a field marked by
unforgettable milestones of daring devotion that
the annals of the land can scarcely equal."**
Frank Shaw, in *Famous Shipwrecks* (1930)

THE story of a lifeboat is inevitably a story of human tragedy, counterbalanced with tales of bravery and the selfless service of its crews, battling against nature's most unpredictable environment. The drama, tension, horror, excitement, and emotional response associated with the saving of lives at sea is focused on the target of their efforts: the Shipwreck.

By giving their services voluntarily to the saving of lives at sea, lifeboat crews willingly enter that hazardous workplace, selflessly going afloat in their attempt to snatch potential victims from certain death. But while striving to save others they also risk becoming victims themselves. Several of the lifeboat stations that appear in this account have experienced great misfortune with the loss of their own lifeboats and crews (specifically St Ives, Padstow, Broughty Ferry, Longhope and our own Penlee station).

So, how can we explain this fascination with maritime disaster that affects many a land-lubber and most mariners? In spite of the fact that there will always be a ghoulish element in any population — those

The wreck of HMS Anson on Loe Bar in 1807, seen here dramatically re-imagined by the maritime artist Geoffrey Huband, inspired the formation of the Lifesaving Apparatus teams.
BY COURTESY OF THE ARTIST

rubbernecks who slow down to gawk at motorway carnage or revel in aviation disasters — other right-thinking, empathetic souls who love ships and boats will be drawn, awe-struck, towards a wreck or stranded vessel which can be seen to function as a symbol of our own mortality and humility.

'A tale of shipwreck seems to have an appeal as unfailing as it is universal for the majority of humankind. [It] appeals to the elemental human instincts of pity, and terror and wonder', writes C. Fox Smith in his book *Adventures and Perils*. The sub-title of his book, which was published in 1936 during the early years of our lifeboat's tenure, reads:

> *Being extracts from the 100-year-old Mariner's Chronicle, and other sources descriptive of Shipwrecks and Adventures at Sea,* which refers to a Marine Insurance Policy document: *'Touching the Adventures and Perils which we the Assurers are contented to bear and to take upon us in this voyage, they are, of the Seas, Men of War, Fire, Enemies, Pirates, Rovers, Thieves, Jettisons, Letters of Mart and Countermart, Surprisals, Takings at Sea, Arrests, Restraints and Detachments of all Kings, Princes and People, of what Nation, Condition or quality soever, Barratry of the Master or Mariners, and of all other Perils, Losses or Misfortunes that have or shall come to the Hurt, Detriment or Damage of the said Goods, Merchandise or ship. . .'*

Of these listed perils, the first is the worst: The Seas.

Wreck watchers

The celebrated novelist John Fowles goes some way to delineate the range of motivations for wreck gazing in his introductory chapter to the book *Shipwreck,* a portfolio of late 19th and early 20th century photographs by the Scillonian Gibson family, which sold many copies to the general

The barque Gunvor was wrecked on the Lizard in 1912 bound Falmouth from Chile, with a cargo of nitrate (guano). The crew climbed ashore off the bowsprit but the cabin boy fell in and was rescued by the Chief Officer

readership when published in the 1970s. Fowles claims the primary emotions can be genuine pity for the victims, the joy of celebrating their successful rescue, and the feeling of intense sadness when they don't survive, although those feelings of pity and fear can in some sense be a cathartic experience. Conversely, the feeling of pleasure at the misfortune of others, '*schadenfreude*', is not uncommon, although a more charitable feeling of 'thanks that it was not me caught up in such a hellish experience' might be more appropriate. The prospect of witnessing selfless acts of heroism on behalf of saviours and salvors is another crowd-puller.

The drama of an exciting spectacle fuelled and augmented by extreme natural forces will have its own attractions, as Fowles claims that 'storms and gales seem to awaken something joyous and excited', especially in those who do not live within sight and sound of the sea. The author goes on to talk about the emotional symbolism of the sea and of ships, bestowing on both the gendered identities which have been attached for centuries. His psychological analysis goes deep, but he doesn't look as far into the visual presence of ships and boats.

The innate beauty of a square-rigged sailing ship is undeniable, but even a squat little coaster or giant 'box ship' container vessel will have an aesthetic appeal to those who self-identify as 'ship lovers'. The bulk and profile of a ship is one aspect, as it cuts through a seascape, making its mark against the organic forms of a natural background. A ship's movement normally appears graceful, purposeful, slow and delicately manoeuvred, particularly in contrast to other forms of mechanised transport (closest, perhaps, to the pace and grace of a working horse). A ship moves in many dimensions — rising, falling and rolling, with pitch, sway, slam, broach and yaw among its motions, always looking as if to break free of its context, especially when embroiled in a raging sea. Once a vessel is stationary, held fast without the lullaby rocking of a swell and grasped firmly in cross-hatched mooring lines, it might amost appear imprisoned.

Even small ships are BIG, especially when they break out of context, closing into one's personal space by appearing on your beach or beneath your cliff-side walk. They are all the more dominating to a small wooden-hulled motor boat rising and falling against a vessel's flank amid the thrashing surf as the sea throws it, within seconds, from below the waterline to the roof of the deckhouse, as happened to the *W&S* when alongside the doomed battleship *Warspite*.

As Frank H. Shaw, author of *Famous Shipwrecks*, wrote during the year when the *W&S* was under construction: 'There is something about a ship, no matter how ugly and utilitarian circumstances have made her, that differs from any other human fabrication, and it is this quality, known by seamen, faintly appreciated by landsmen, maybe, that makes of a wreck a deep and stunning tragedy.'

Wreckers, Salvors and Saviours: Facts and Fictions

It would be disingenuous to discuss the story of saving life along the Cornish coast without referring back to the mythology of the 'wreckers'. There does exist abundant testimony from contemporary 18th and 19th century sources describing the activities of those who profited from the misfortunes of mariners. The artist, author and explorer Frederick Whymper, in his book *The Sea: Its Stirring Story of Adventure, Peril, & Heroism,* published in 1885, relates shocking tales of inhuman, even barbarous, behaviour including luring ships onto rocks with false signal lights; shadowing doomed or disabled vessels (like perversely inverted pilots); plundering beached vessels of all cargo, personal possessions, ship's equipment and fittings; murdering survivors, and even mindlessly setting fire to salvable hulks. On occasion, hundreds, sometimes thousands, of locals (described in many period documents as 'the Country') would gather on cliff tops, beaches and in caves waiting to participate in the pillage.

The Victorians' love of melodrama extended to the dreaded practice of the 'wreckers', since shown to have been as much myth as reality

But it was not only the dispossessed or impoverished classes who indulged in what some had described as 'sport' as well as a business. The 'gentry' and landowners of the likes of the Killigrew dynasty, were able to justify the seizing of cargoes as being 'proprietors of the rights to wreck'. Revenue officers, magistrates and clergymen were also sometimes implicated, as exemplified in the phrase 'brandy for the parson'. In one, often repeated and possibly apocryphal, incident a wreck was announced during a church service and the vicar begged the

congregation to give him time to remove his cassock so they could 'all start fair' in the race to reap the rewards of wreck. In the Isles of Scilly there is a prayer asking for there to be no shipwrecks; 'but if they do happen, please God let them happen here'.

Historically, wrecked property belonged to the king, until Henry1 (1100-1135) decreed that 'neither wreck nor cargo should become the property of the Crown if any man of the crew escaped with life to shore'. However, as this could have been construed as an invitation to murder, the Act was later amended to include 'any beast'. Richard 1 (1189-1199) added that wreck should only belong to the king when neither an owner nor heir could be found for it. Furthermore, he decried the 'accursed custom by which a proportion of any wreck would belong to the lord of the manor where the wrecks take place, and that pilots, for profit, do purposely run ships under their care upon the rocks [. . .] The law declares that all false pilots shall suffer a rigorous and merciless death, and be hung on high gibbets, while the wicked lords are to be tied to a post in the middle of their own houses, which shall be set on fire at all four corners, and burnt, with all that shall be therein.' It was added that 'if people, more barbarous, cruel and inhuman than mad dogs, murdered shipwreck folk, they were to be plunged into the sea until half dead, and then drawn out and stoned to death'. Centuries later, George ll (1727-1760) passed an Act condemning wrecking as a disgrace to the nation, and that 'death should be the punishment for hanging out false lights; killing shipwrecked persons; stealing cargo or

"Those who murdered shipwrecked folk should be plunged into the sea until half dead, then drawn out and stoned to death

wreckage, whether anyone on board remained alive or not.'

In fact, according to accounts quoted in *Cornish Wrecking, 1700-1860: Reality and Popular Myth,* there was only one recorded instance of a Cornish person being executed for wrecking under the Act of George II, and that was for a comparatively petty offence committed by a very elderly person. The author Cathryn J. Pearce refers to an article in the *Sherborne Mercury* [which covered Westcountry news between 1748 and 1867] stating that in 1767, Wm Pearce was executed at St Stephens, near Launceston, believing to the last that he was only acting according to custom. His sentence was passed as a deterrent to others. He was 80 years old and had been found guilty of taking an 'inconsiderable quantity of cotton from a wreck'.

However, when the Portuguese carrack (galleon) *St Anthony* went ashore at Gunwalloe, Mount's Bay in 1526 with the loss of many crew, accounts state that the survivors accused some of the local gentry of robbery with violence. The ship, belonging to King John III of Portugal, carried cargo including copper, silver, precious stones, tapestry, cloth, brass items, musical instruments and other valuable artefacts. Some of these items have been retrieved and are on display at various sites, including those at the Shipwreck Musuem in Charlestown.

While the Cornish wreckers have been so mythologised, it should be recognised that the pursuits of piracy, smuggling and wrecking were practised on every coast in the land — on the East Coast of England and Scotland and the Thames Estuary as much as the South-West. But as the South-Western counties were in so many ways detached from the seats of power in London, as communications and law enforcement were so weak, and with so many opportunities to practice their occupation, Cornish wreckers did catch the imagination. In December 1846, the *Penzance*

Gazette commented on the term 'Cornish Wreckers':

'Whenever a wreck takes place, plunder generally ensues, be the locality where it may — and the people of the 'neighourhood' are spoken of as 'having very improperly conducted themselves in carrying off the property, &c, &c'; but if it happen on the coast of our County, they are at once held up to execration as 'Cornish Wreckers'. We cannot submit to the imputation that the shipwrecked mariner is worse treated in Cornwall than he would have been had his misfortune befallen him in Devonshire or Somerset, on the coasts of Scotland or Ireland, or at the mouth of the Thames itself.'

One example given in the article to show the Cornish people's innate humanity is the story of the brig *Neptunus* on passage from Cadiz to Christiansand which was compelled to run aground near Land's End after springing a leak in November 1846. The first boat to come to their aid was the *Briton* of Mousehole, which took off the crew's chests, clothes and bedding for safekeeping, and for which they claimed no salvage. Other boats then arrived and cleared all the remaining materials. In this service, the Coastguard united with the crews of the Bay boats in salvage and protection; and so efficiently was this operation carried out that 'not a single item was later found to be missing'.

The *Gazette* recounts the tale of the Norwegian schooner brig *Elizabeth* which came ashore at Gunwalloe, eleven days later. The ship quickly became a total loss, with the deaths of two men and a boy. 'And such would have been the fate of them all, but for the united exertions of the crowd assembled on the beach, who succeeded at the imminent risk of their own lives, and with considerable injury to their limbs, in saving the master, mate, and two seamen. These unfortunates had stripped themselves nearly naked to have a better chance of getting on shore; but their rescuers sent to their own houses for dry clothes which they [survivors] wore till new ones could be provided.' The *Gazette* continued its re-assessment of Cornish morality by referring to an arrangement respecting the western part of Cornwall whereby those who rendered real service in the case of need at wrecks were liberally and promptly rewarded, whilst those who indulged in plunder were visited with 'condign punishment'. The credit for

this initiative was given to Richard Pearce, a Lloyd's agent at Penzance, whose experience and campaigning energy over three decades helped moderate the crime of wrecking in the county and bring about reform in the treatment of shipwrecked seamen.

The takings from plundered wrecks were often substantial. During the 18th century, European merchantmen returning from the East and West Indies and the Americas were loaded with riches, which cargoes had sometimes been looted at source or en route. Trans-European trade was also dynamic. Among the manifests of vessels whose wreck was recorded there was frequent mention of multiple hogsheads (huge barrels) of wine, brandy or rum; bales of silk; animal hides; fine china and silverware; spices and exotic fruits; consumables such as coffee, sugar, salt and tobacco; family and dynastic heirlooms and spoils of war.

At that time many of these imported treasures would have been the profits of the slave trade, *aka* the Black Triangle, a three-way traffic in which manufactured goods were exported to the Colonies from Europe, traded for enslaved Africans who were transported to the Americas in brutally inhumane conditions with great loss of life, never to return, and exchanged for high-value imports to be brought back from the New World. Much of Britain's inherited wealth has recently been shown to have accrued from the profits of slavery and the compensation paid to those traders and slave owners when the practice was finally made illegal. Disconcertingly, it has come to light that several of the early ship owners and entrepreneurs who helped inaugurate the lifeboat service were profiting from that nefarious trade, and even English heroes such as Sir Francis Drake and Lord Horatio Nelson, no less, have also seen their repututaions suffer from re-examination.

In 1589 Sir John Killigrew, a notorious pirate and landowner of Southern Cornwall described one cargo of goods 'cast awaye' in Mount's Bay valued at £100,000 which was an astronomical amount in those days (roughly equivalent to some £17 million in 2018.) However, not all cargoes were so exotic or valuable: a splendid full-rigged ship carrying a load of grand pianos was wrecked in Cornish waters around the same time as another three-master which was loaded with the most prosaic 'dirty' cargo of slag from the South Wales coalfields.

The practice of pillaging or scavenging, if not exactly wrecking, did revisit the Westcountry in 2007 when a 275-metre (900-foot) container ship of 60,000dwt started to break up 50 miles south of the The Lizard, but it was the residents of neighbouring Devon, rather than Cornwall, who stood to profit from its misfortune. The crew of the *MSC Napoli* had abandoned ship and were rescued by helicopter from RNAS Culdrose. The ship could have been taken to Falmouth but was eventually towed along the coast of Cornwall to Lyme Bay, where it was deliberately beached on the Jurassic Coast at Branscombe, Devon. Once some of its many containers had started to wash off its deck the beach was invaded by dozens, even hundreds, of treasure hunters who succeeded in carrying away a huge catalogue of goods, from brand new high-end BMW motorcycles to empty wine barrels and household belongings. The Receiver of Wreck eventually moved in to stop the plunder, but not before many of the containers had been pillaged. Months later, those salvagers who had reported their finds to the Receiver were told they could keep what they had found.

And then, of course there were the pirates. Again, piracy has been widespread on the high seas for millennia — probably ever since ships carrying goods strayed too far from safe haven. Buccaneers were pirates, plain and simple, while privateers were legally sanctioned for generations, on ships which alternated between government and private ownership, whichever was the most profitable at the time. Cornwall does have a special case to answer in this regard, as the name 'Pirate' is attached to so many of the county's activities, not least the official brand name of their rugby team, local radio station and so on. And there is also the well-known light opera associated with the town of Penzance. Historians have long been aware that the good people of Mousehole have not always been fisherfolk, but in bygone days they were more usually earning their livelihoods from looting and 'sacking' of vessels gone astray. Not for nothing are they known as 'cut-throats'.

One early record of a Mousehole wreck concerns a large Spanish galleon that ran onto Low Lee Ledge off Penlee Point in January 1635. Sir Francis Basset, the Vice-Admiral of the Western shore, salvaged a shipment of hides to be taken away the next day 'but the rebels of Mousehole came in their boats in the night and carried them all away to their homes'. Homeward bound from the Indies, the galleon had already been captured and looted by the Dutch. Putting into Gwavas Lake off Newlyn, she hit the Low Lee Ledge. Attempts at salvage were opposed by the inhabitants of Mousehole who raided the ship at night and took away '200 hides'. A looted cannon from this ship was salvaged by the *Greencastle* in 1916 and for many years stood in front of Penzance library before being stolen [again].

Lifesavers on shore and sea

Henry Trengrouse is not a name that has resonated as long as it should, but he was one of West Cornwall's technological and social visionaries, along with his contemporaries Humphry Davy, (Chemist and Inventor) and Richard Trevithick (Steam Engineer). A resident of Helston, Trengrouse was present at one of Mount's Bay's most monstrous shipwrecks when the frigate HMS *Anson* was driven ashore at Loe Bar on Christmas Eve 1807, resulting in the loss of more than 100 lives. That event left a lasting and nagging impression on the man who witnessed so many deaths of people who took the wrong decision in trying to escape their doom among the breakers. He observed that most of those fatalities could have been prevented if the crew had stayed aboard and waited for the tide to ebb, rather than leaping into the roiling sea. Although Trengrouse had already attempted to design a lifeboat, he was inspired to attack the problem from the shore rather than sea, and he developed an improvement on the system already proposed by Captain George Manby

SHIPWRECK INVESTIGATED,

FOR THE CAUSE

Of the great loss of lives with which it is frequently attended;

AND

A REMEDY PROVIDED,

IN A PORTABLE AND PRACTICABLE

LIFE PRESERVING APPARATUS;

Which is calculated for, and is necessary to become, a part of every ship's equipment.

Its Efficacy is also Proved,

By demonstrative application to the cases of a great number of recent Shipwrecks has been narrated; from which it is evident, that at sight of them only, it might have been rendered instrumental in preserving

NEAR TWO THOUSAND LIVES.

A great number of the narrations of Shipwrecks are very interesting, and many of them have never been before published.

By HENRY TRENGROUSE,

HELSTON.

for shooting a line to stranded vessels with which to haul survivors ashore. A messenger line was fired at the grounded vessel carrying a heavier hawser with a bosun's chair attached. Manby used a mortar to fire the device, but Trengrouse proposed using a rocket that could be fired from a musket. His device was lighter than the mortar, and the line was less likely to break because the acceleration was more gradual. It was also much more portable for use on shore and could easily be carried aboard ship. It first came into use in 1808, just one year after the *Anson* disaster. In 1817 Trengrouse published his seminal work with the full title of:

> '*Shipwreck Investigated* for the cause of the great loss of lives with which it is frequently attended and a remedy provided in a portable and practicable life saving apparatus; which is calculated for, and is necessary to become, a part of every ship's equipment. Its Efficacy is also Proved, By demonstrative application to the cases of a great number of recent Shipwrecks herein narrated; from which it is evident, that at eight of them only, it might have been rendered instrumental in preserving NEAR TWO THOUSAND LIVES. A great number of the narratives are very interesting, and many of them have never been before published.'

Ten years after his book's publication the Royal Navy, and then Trinity House, recommended the use of his apparatus and although he received little financial credit, Trengrouse's device came into common usage and led to the formation of the Life Saving Apparatus (LSA) an association which saved many lives including some from the most infamous Cornish shipwrecks, and several from shouts attended to by the *W&S*, the subject of our book. Locally, there were LSA stations at Mousehole, Porthleven and The Lizard.

Putting to sea at the service of others

"The coastwise wreck is in itself a terrible disaster," writes Frank Shaw, "but thanks to the lifeboat service [. . .] the sea's grim toll of human life has been heroically lessened in the shoal and ragged waters of the shores of the world."

To most people, ships are witnessed at a distance; only those aboard the casualty or rescue vessel really *know* what is happening out there, and wrecks within sight of land are experienced at two or three stages removed. Thanks to the testimonies of several lifeboat crew members and survivors of shipwreck quoted in these pages, we

The Elizabeth and Blanche II *(to right of picture) alongside the 4,000 ton collier* Cragoswald *which ran aground on Low Lee Ledge, Mousehole in 1911, with a crew of 27, all of whom were rescued by the lifeboat .The ship was later refloated and repaired*

NIM BAWDEN COLLECTION

have some idea of the what it is like to be involved in such a calamitous event. And whatever misfortunes might befall those on solid ground they should always remember the adage that 'Worse things happen at sea'.

Even within sight of land, a foundering ship will be out of reach, its suffering unimaginable; the witness experience will be vicarious and often the only point of

RICHARDS BROS

The sad sight of the Umbre, *wrecked off Pendeen in 1899 on passage Liverpool to Amsterdam. The crew of 13 were rescued by the Brixham trawler* Evangelist *and landed at Newlyn*

connection will be the lifeboat, which the witness might well have visited in its boathouse, or watched bobbing at its harbour mooring. There the lifeboat might look impressive in size and power, but once alongside a looming casualty vessel that boat will look and feel insignificant, dwarfed and threatened by the lurking hulk. Wrecked ships and boats are intensely sad, and even the smallest boat looks forlorn and incongruous when it's turned upside down on the sea shore: a boat should be afloat. Those casualty ships are the victims and anti-heroes of this story, whose actual hero is our boat, built of wood by old-school craftsmen; the kind of vessel that inspires respect, admiration and even love in the hearts of true boat lovers.

For practical or sentimental reasons, the lifeboat crew will save a boat when they can, but their primary objective is to save a life, whatever the circumstances. The crews are what we now call 'first responders' and their paramedic skills are as important as their boat handling. Apart from the obvious physical dangers expected during a shipwreck, lifeboat crews can be faced with industrial injuries aboard ship, as well as a whole range of sickness and disease, often undiagnosed. When they take out a doctor, ambulance team or paramedic for a 'medevac' (medical evacuation) of a suffering seafarer they will expect to be faced with the tricky procedure of manhandling a stretcher over the rails of a high-sided vessel and onto the pitching, heaving deck of their own boat, with its limited deck space.

The first shout for the *W&S* exemplifies the extremes of the lifeboat ethic. The saving of a survivor will be seen as a success, and must impart a sense of triumph. Conversely, the landing of a fatality from a wreck can be regarded as a sombre acknowledgement of (or even a sacrifice to) the power of the sea, rather than a failure, but it will inevitably be felt by the crew as a loss.

One of the paradoxes of life-saving is that the saviours are most usually involved after the event, to 'pick up the pieces'. The lifeboat call-outs might occasionally come in advance of an incident, but are most likely to be received during the height of a crisis or, more frequently, after the worst has happened. In the days before seafarers were able to enjoy the benefits of electronic communications it was most likely that lifeboats responded once a drama of some sort had started to unfold — often some time after an incident that might have occurred without witnesses. Then again, there are occasions when nearby vessels close to the casualty have been in a position to answer the legal and moral obligations of all mariners to come to the aid of vessels in distress, as happened during the first *W&S* service to the steamship *Opal*.

During the days of sail, such vessels that were known to have been wrecked around Cornwall were most often coastal traders visiting local ports, or merchantmen on oceanic voyages to and from the economic capitals of Europe. Frequently, ships listed as 'overdue' at their destinations had simply disappeared and no sight nor sound was ever received. An additional danger to seafarers was the fact that most of the navigation charts in use at the start of the 20th century were generally less than precise, and had been prepared by surveyors working aboard sailing vessels using equipment that had become obsolete. Overloaded cargo ships had cost the lives of thousands of mariners, until the Merchant Shipping Act of 1875 brought the first regulation to the business. Its sponsor Samuel Plimsoll has been immortalised in the term Plimsoll Line, the load line mark on a ship's hull which indicates if it has sufficient freeboard, and which was adopted internationally in 1930.

At the beginning of the era of steam, around the turn of the 19th/ 20th centuries, new trading patterns were established; most significant to our story was the export of Welsh coal from Bristol Channel ports and the extension of tramp steaming. The technology to build and operate the new generation of iron-hulled and steam powered vessels was still being developed. Conditions for seafarers were grim, with poor communications and equipment failure adding to their woes. Masters, officers, engineers and crews were often less than competent to operate the inefficient machinery, and crew accommodation and conditions could be even worse than those aboard the sailing vessels. In the years of recession and economic depression, between the two World Wars, there were suspicions that the disappearance of many ships was often to the benefit of corrupt ship owners, leading to tales of 'coffin ships', or death ships: vessels so manifestly unseaworthy that the owners might plan to scuttle them for the insurance money.

Technological developments

One of the most significant technological developments to transform maritime communications was the invention of wireless or radio-telegraphy. Since the 1830s, telegraph lines on land had provided a linear message system via overhead wires supported on telegraph poles. An operator would tap on a key, creating pulses of current to spell out a message in Morse code, a system of dots and dashes which was already in use for signalling by flashing lights. The most well-known message, which has passed into the lexicon of common usage is the universal plea SOS, represented by the Morse code • • • — — — • • • and meaning Save Our Souls, which was adopted internationally in 1905. In the second half of the 19th century telegraph networks covered most industrialised nations, and eventually submarine cables carried messages between the continents. The first two of these cables came into Mousehole and operated until 1929, and 14 of those that eventually reached out to British Empire territories actually still come ashore at Porthcurno. The first maritime comms carrying these messages were received at land stations, such as the one established by Lloyd's on the Lizard, which were then transmitted to and from vessels at sea by flag and light signals. To speed up communications from this long winded procedure shorthand codes were drawn up and radio operators would carry dictionaries of signal codes.

The discovery of radio waves in 1887 led the Italian inventor Guglielmo Marconi to the development of radiotelegraph transmitters and receivers by the end of the 19th century, which meant that spoken

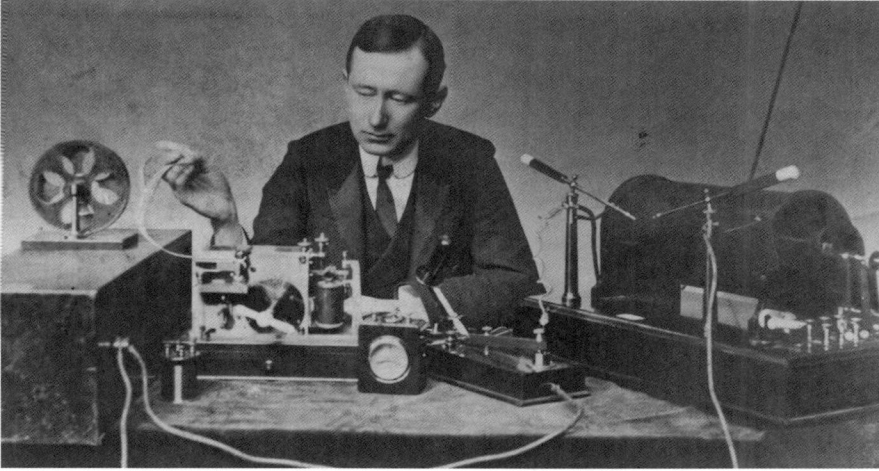

Guglielmo Marconi discovered radio waves in 1887 and went on to develop a system of communication which continues to this day

messages could now be sent and received. Marconi also developed the echo sounder, and by 1936 his Echometer was being fitted to British ships at the rate of one per day. The international distress call 'Mayday' was introduced in 1923 and accepted internationally in 1948. It had been proposed by a radio officer at Croydon Airport in London, who chose it as an anglified version of the French *m'aider*, meaning 'help me'. By January 1944, the Royal Navy was testing the Decca Navigator system and the first commercially available radar devices became available after the war in 1947

The idea of a weather 'forecast' came about in the late 19th century through the efforts of Admiral Robert FitzRoy, renowned as the Captain of Charles Darwin's HMS *Beagle*. In the years between 1855 and 1860, 7,402 ships were reportedly wrecked off the coasts of Britain with a similar total of lives lost. Shortly after, in 1860 FitzRoy started issuing storm warnings via the new medium of telegraphy. Weather data, gathered locally from the coast, would be sent to his central office in London and forwarded to relevant harbours where a storm signal would be hoisted. Soon afterwards FitzRoy's forecasts were being published in *The Times* newspaper, and his descriptions 'fine', 'fair', 'rainy' and 'stormy' came into standard usage. When the BBC started radio broadcasting in 1922, followed by national coverage in 1925, shipping forecasts were among the first programmes to be broadcast daily. As the technology improved after World War 2, ocean-going ships were more able to fend for themselves, and the next generation of seafarers to benefit from the technological improvements, and especially weather forecasts, were the fishermen, of whom there were many working in Cornish waters and whose vessels were the most common type of casualty during the 1950s. As the fishing fleets declined in number they made fewer calls on the lifeboat's services and, going into the 1960s, there was an increase in calls from leisure craft, which started to take up more of the RNLI's time, always allowing for the occasional call out to small merchant vessels in distress.

All these technological developments made the mariner's job that much easier, more predictable and generally safer, notwithstanding the inherent dangers of over dependence on systems and the extra administrative burden, but the sea is a fickle and changeable environment and human error is just a part of human nature.

The Build of the Boat

Watson and White's: Proud parents

A REGULAR biography would begin its narrative with the parental
origins of its subject. A person would have a family tree, a dog
its pedigree, a horse its stud record, whereas a boat is credited
to its designer and builder. The subject of this 'nautobiography' the
W&S (ON736), has unarguably high-class parentage: designed by the
successors of George Lennox Watson (1851–1904) and built in the yard
of J. Samuel White. Place of birth Cowes, Isle of Wight; date of delivery
February 26, 1931; weight at launch 11 tons.

The Isle of Wight had been the base of shipyards that built
vessels for the Royal Navy since around the start of the 19th Century. In
1811, Thomas White, a third-generation member of the White family of
boatbuilders from Broadstairs, Kent, purchased the old Nye shipyard at
East Cowes, looking to expand the company's business. The firm later
developed other yards in West and East Cowes on the Medina river, just
south of the Chain Ferry. Along with several other shipyards, notably
S. E. Saunders and Groves & Gutteridge, Cowes became a major supplier
to the Royal Navy of smaller ships, up to the size of destroyers, a trade
which J. S. White & Co. Ltd augmented with construction of a wide range

*J.S. White's
workshop in the
1950s where yard
number 1705
was built 20 years
previously*
RNLI PHOTO

of vessels from ship's lifeboats to ferries, ocean liners and private yachts, both sail and powered.

The company earned an enviable reputation for craftsmanship and building to deadlines and budget, and at various times employed anywhere from 500 to 1,000 designers, shipwrights, metal workers, fitters and other specialist tradesmen.

During the 1840s White's collaborated with a local designer, Andrew Lamb, to build a patented self-righting vessel to be known as the Lamb and White Lifeboat. To build these boats a dedicated lifeboat construction workshop was established at the West Cowes yard. Designed primarily for use on board ships, these were whaler type open boats of carvel construction (where the planks are laid flush and smooth, as opposed to clinker-built where the planks overlap each other), with a series of air-tight compartments along each side and at both ends, which made them less likely to capsize than existing boat designs. These boats came in a range of sizes from 27ft to 32ft and were soon taken up by several shore-based lifesaving organisations, which eventually included the National Shipwreck Institution (NSI), the forerunner of the RNLI. Hundreds of the boats were built to serve on naval and passenger ships, as well as at shore stations, where they would have been kept on horse-drawn carriages to be launched from a beach. When the Coastguard Service was absorbed into the Admiralty in 1860, more than 100 whalers stationed around the coast were replaced by 27ft versions of the L&W whaler/lifeboat. Soon after, the Admiralty decreed that all Royal Navy ships should be so equipped and an order was placed for 500 Lamb & White lifeboats.

The majority of lifeboats in service with the RNLI during the 19th century had been of the self-righting type of pulling (rowing) and sailing boats, which were successful at righting themselves after a capsize but their shallow draft design compromised on general stability which came to be considered more important: better not to capsize in the first place. To counteract that shortcoming the naval architect George Lennox Watson conceived the displacement hull type that bore his name: a name that resounds into the 21st Century. Watson had concluded the self-righting properties were useful for smaller pulling boats designed for surf and inshore work, but for longer-range offshore use, non-self-righting boats could be faster, more easily manoeuvred and more seaworthy.

In 1888 the first Watson class lifeboat, the 42ft 3in long *Edith and Annie* (ON208), was built for the RNLI, and during the next 27 years, 42 models of the Watson-class pulling-and-sailing types were built in a variety of lengths, the most common being the 38 ft model. Watson lived long enough to advise the Institution on the application of steam and internal combustion power to the larger lifeboats, although he died before the full trials of motor lifeboats were completed.

Fanatical attention to detail

George Lennox Watson (1851-1904) was born in Glasgow, the son of a doctor and grandson of Sir Timothy Burstall, engineer and builder of early experimental steam carriages, including the *Perseverance* which was entered, and earned an award, at the 1829 Rainhill Trials, won by the Stephenson brothers' *Rocket*. George Watson grew up in the heart of Clydeside industry at a time when it was reaching its peak. As a boy, he would holiday at his family's house at Inverkip on the Clyde estuary in Ayrshire. From there he could see the ships and yachts sailing out on the estuary and this sparked his interest in the world of yachting. During his holidays he befriended a local yacht hand called William Mackie and, due to his youthful enthusiasm and passion for yachts, he was soon telling his friend how he would go about designing a yacht. As soon as he was able, Watson served his apprenticeship with the Clydeside ship builders and engineering firm, Robert Napier & Son. This equipped Watson with all the skills he needed and, in 1873, at the age of just 22 he set up the world's first dedicated yacht design office.

His early success in the racing classes of the day soon expanded his client base and brought his name to the fore as the most innovative designer of the day. His designs became iconic and many were involved in some of the most notable moments in maritime history.

Watson remained dedicated to his work and exercised almost fanatical personal attention to detailed aspects of his yachts, both in design and build, before his early death in November 1904, aged only 54. At the time of his death Watson had over 400 designs to his name. He had successfully mentored his apprentice J.R. Barnett to become an accomplished designer in his own right, and Barnett was elevated to the post of Managing Director and sole partner. Barnett carried forward Watson's design legacy, which continued through subsequent directors of the firm that today still bears the name of its founder. In the 21st Century, J. L. Watson & Co. Ltd, now based in Liverpool, continues to design, build and refit superyachts of all classes.

While learning his trade at Napier, Watson had become interested in developing theories of hydrodynamics in the design of racing yachts. He moved on to work at J.&A. Inglis, Shipbuilders, where he founded the world's first yacht design office dedicated to small craft. His first design, an 8-ton cutter, *Peg Woffington*, featured an unorthodox reverse bow. His business developed quickly, with commissions to design steam and sailing yachts for wealthy industrialists including the Coates family, the Vanderbilts, the Rothschilds, the Earl of Dunraven, Sir Thomas Lipton, and royalty such as Wilhelm II, the German Emperor and Albert Edward, Prince of Wales. There

Watson's most celebrated design was the Royal Yacht Britannia *built in 1893, which became the most succesful racing yacht, raced by Edward VII as Prince of Wales and taken over by his son King George V, seen here at the helm in 1924*
COURTESY G.L. WATSON

were commissions for four America's Cup challengers and the A-class schooner, *Rainbow*, the largest schooner of its time, built in 1898.

Watson's most famous design was HMY *Britannia*, built in 1893 for the Prince of Wales, later Edward VII, who raced her successfully before she was handed over to his son King George V in 1910. In a 43-year career *Britannia* became the most successful racing yacht of all time. In the Mount's Bay Regatta of July 28, 1894 the America's Cup defender *Vigilant* was piloted by Benjamin Nicholls, and the *Britannia* piloted by Ben's brother Philip Nicholls, both of whom were Trinity House pilots working out of Penzance. Crowds of people came to west Cornwall from far and wide to watch *Britannia* win the race by just over seven minutes. Philip Nicholls later advised the company on the design of *Elizabeth and Blanche II*, delivered to Penzance in 1899, which was the first lifeboat to be housed in the newly-built Penlee boathouse in 1913.

Concurrent with his work on prestigious sailing and motor-powered yacht designs, Watson continued working on the development of lifeboats and in 1887 he became chief consulting Naval Architect to the RNLI, a position which his company, G.L. Watson & Co. would maintain through several generations into the late 1960s.

In the 1830s Sir William Hillary, founder of the RNLI, had proposed a steam-driven lifeboat, but this was not realised until 60 years later when the first steam-and-sail-driven lifeboat the *City of Glasgow*, went into service in September 1890 at Harwich Lifeboat Station (where

our boat ON736 would arrive some 125 years later). That hydraulic, steam-powered lifeboat was driven by waterjets rather than propellers. A second boat of the same name, built by J.S. White, introduced in 1901 and sold out of service in 1917, was the RNLI's last steam boat. Five steam lifeboats of that type were built, with three of them being propeller-driven. Between them, these lifeboats were in service for more

The second steam-powered City of Glasgow was stationed at Harwich between 1901 and 1917

than 40 years and saved 570 lives. But they were restricted to deep-water moorings, and could take a long time to fire up when needed in an emergency. Subsequently lighter, faster and more practical petrol-driven boats would soon come into use, still using the Watson designed propeller 'tunnel'. When the *James Stephens No4* was wrecked at Padstow in 1900, with the loss of eight crew, it signalled the end of that era, and no more steam-engined RNLI lifeboats were ever built.

At the start of the 20th century, industry was undergoing a mechanical revolution, and in 1904, the year of George Watson's death, the RNLI fitted the first petrol engine to one of the Watson 38ft self-righters. After several experiments by the RNLI on various adapted non-powered craft, which were not Watsons, the first new-build motor lifeboats were constructed in 1908, two of which were Watson types.

Following the era of the steam-driven boats, a series of Watson motor lifeboats in various classes was built over the next 55 years. There were 11 lengths of boat in eight separate classes, with variations to meet the requirements of different stations.

Between 1908 and 1930 there were 15 boats of 38ft, 40ft and 40ft 6 in. From 1912 to 1925, there were 22 examples of the 45ft class. Between 1926 and 1935 there followed 23 Cabin class boats of our 45ft 6in type. Another 13 of the 41ft boats were delivered between 1933 and 1952. The biggest production runs were 28 models of the 46ft boats, built between 1936 and 1946, and the same number of 46ft 9in versions over the next 10 years. These were followed by 10 of the 42ft boats, and a final run of 18 boats in the 47ft class, which saw the last of the Watson lifeboats delivered in 1963.

The 45ft 6in Watsons marked an important stage in the evolution of RNLI boats, being the first to feature a cabin with a shelter to provide some protection from the weather for crews and survivors, and boats in this series were also the first to be powered by twin engines. The *W&S* was of standard Watson Cabin class design, 45 feet 6 inches in length, although there were some modifications made, and our boat was actually half an inch shorter than specified. Unlike previous Watsons, this class of boat was flush-decked with no end boxes. The 23 models of this class of boat were built at three different yards in Cowes, Isle of Wight (and one in mainland Scotland) between 1926 and 1935. According to the RNLI, in 1927 there were reportedly eight boats of this class scheduled to be built that were awaiting donors.

The first two boats, ON698 and ON699, were single-engined models built at S. E. Saunders in 1926 and the next two, ON700 and ON701, powered by twin engines, were built by J. Samuel White in 1927. Saunders built five, and another two in 1930 when the company had become Saunders Roe (later famous for its flying boats and hovercraft). Fourteen of the boats, including ON736, were built by J. Samuel White and one by Groves & Gutteridge. The odd boat not built at Cowes, was ON774, *Charlotte Elizabeth*, built by Alexander Roberston at Sandbanks, Argyle in 1935, becoming the first motor-powered lifeboat launched in

Thomas McCunn was the last of the Watson Cabin-class 45ft 6in boats to see service, which included a spell as relief boat at Penlee

The 45 ft. 6 in. Watson (Cabin) Motor Life-boat.

Scotland. Most of this series of Watsons served for several decades on RNLI stations and ON759, *Thomas McCunn*, built at Cowes by Groves & Gutteridge was the last to serve, being operational until 1972 and including a period as relief boat at Penlee. That boat is now preserved at Longhope in Orkney.

ON698 and ON699 were powered by the same 80bhp Weyburn DE6 6-cylinder petrol engine as the previous single-engined Watsons, while the third boat (ON700) was the first lifeboat to be fitted with two power units driving twin propellers. The engines used in the twin-engined boats were smaller 4-cylinder units from the same maker, each rated at 40bhp, providing the same total power output of 80bhp. These were the engines subsequently fitted to 'our' boat the *W&S*. Following one unsuccessful experiment gearing the twin-engines to drive a single screw (ON701), from ON707 onwards all RNLI lifeboats were to be twin-engined. The drop keel fitted to the single engine boats was deleted from the twins.

The *Life-Boat* Journal of March, 1933 gave a concise assessment of the boat and its significance to the development of motor lifeboats: 'The 45-feet 6-inches Watson (Cabin) Motor lifeboat was the first type of Motor lifeboat in the Institution's fleet to be provided with a cabin, and the first of the type was built in 1923. This lifeboat is a development of the famous Watson lifeboat, which was designed by the late Mr. G. L. Watson, of the Glasgow firm of yacht builders, Messrs. G. L. Watson & Co., who was appointed the Institution's Consulting Naval Architect in 1887. In 1890 Mr. Watson designed two Lifeboats, one 43 feet by 12 feet 8 inches, weighing nearly 11 tons,

S.G.E. (ON804) was launched at White's Cowes yard in 1938. The boat served nearby at Yarmouth Isle of Wight and in the relief fleet
RNLI PHOTO

which was a sailing lifeboat, and the other 38 feet by 9 feet 4 inches, weighing 4 tons 14 cwt., which was a pulling lifeboat. These two boats were the fore-runners of the stable, non-self-righting Lifeboat of today. The Watson (cabin) type was, as first designed, 45 feet by 12 feet 6 inches, but in 1926 she was lengthened by 6 inches, in order to get an easier run on the water-line and a little more speed. Her displacement in service conditions is 20 tons, and she has a mean draft of 3 ft 8 inches. She is divided into seven water-tight compartments, and is fitted with 142 aircases and ten relieving scuppers. She has a forward and an after cockpit, both fitted with shelters, with room in them for twelve people, and a cabin which will take twenty people, with sitting accommodation for ten. The cabin is fitted with an electric fan, which can be used to ventilate the hold as well as the cabin. In rough weather she can take ninety-five people on board. She is built with a double skin of mahogany, keel of teak, ribs of Canadian rock-elm and stem and stern posts of English oak. She is driven by two four-cylinder 40 h.p. engines. They are in a watertight compartment, and each engine is itself water-tight, so that they would continue running even were the engine room flooded and the engines themselves entirely submerged, for the air-intakes are well above the water-line, even when the boat herself is water-logged. The exhausts are carried up a funnel amidships.

Her maximum speed is 8¼ knots, which, having regard to her speed-length ratio is equivalent to a speed of nearly 34 knots in a vessel of the size of the *Mauretania*. As with all the Institution's Motor lifeboats there is a great reserve of power, so that the maximum speed can be maintained even in very severe weather. This type carries 108 gallons of petrol and the engines' consumption is 7½ gallons an hour at full speed, so that she can travel 116 miles at full speed without refuelling. She carries a staysail and trysail, which can be used either with the engine running or as auxiliary power in the event of any failure on the part of the engines. She carries a crew of eight men, has a line-throwing gun, an electric searchlight and a mechanical capstan, is lighted throughout with electricity, and has a fire-extinguishing plant, worked from the deck, which can throw jets of Pyrene fluid to all vital parts of the boat.

There were 29 Watson (Cabin) Motor lifeboats [of varying lengths] in the Institution's fleet, stationed at Holy Island (Northumberland), Teesmouth (Yorkshire), Cromer (Norfolk). Margate (Kent), Yarmouth (Isle of Wight), Porthdinllaen (Carnarvonshire), Douglas (Isle of Man), Portrush (Antrim), Dunmore East (Waterford), Humber (Yorkshire), Clacton (Essex), Newhaven (Sussex), Selsey (Sussex), Fowey (Cornwall), Penlee (Cornwall), St. Mary's (Isles of Scilly), Tenby (Pembrokeshire), Angle (Pembrokeshire), Fishguard (Pembrokeshire). Piel (Barrow, Lancashire), Longhope (Orkneys), Aith (Shetlands), Thurso (Caithnessshire), Cromarty (Cromartyshire), Montrose (Angus), Dunbar (Haddingtonshire), Donaghadee (Co. Down), Rosslare Harbour (Co. Wexford), and Courtmacsherry (Co. Cork). The original lifeboat of this type, stationed at Tenby, was placed in the Institution's reserve fleet in 1933.'

During their service, 45ft 6in Watsons launched on service 2,587 times and are credited with saving 2,613 lives. The single biggest contributor being the Humber lifeboat *City of Bradford II* (ON 709), which in twenty five years at the station launched on service 228 times, saving 305 lives.

The Birth of the Boat

'Built by Craftsmen for Craftsmen'

THE keel of our boat, yard number W1705, was laid in the lifeboat workshop at J. S. White's yard in late summer 1930. In the 21st Century we would describe the build process as traditional craftsmanship, but back at the start of the previous century, White's ship and boat-building was considered a state-of-the-art industry. The yard was fitted with the latest steam-driven and mechanised equipment, such as lathes, drills, hoists, cranes, winches and other machinery. Nevertheless, the shipwrights in the lifeboat sheds were craftsmen who used traditional boat-building hand tools such as adzes, drawknives, beetles (mallets), caulking irons, roving irons and horning poles, as well as the more familiar saws, chisels and hammers. The procedure was far from mass production, with only two boats of this class built by each of two ship-builders each year through the 1920s and Thirties. Two instructional films showing the process were produced by the RNLI in the 1950s for cinema audiences, to promote the organisation's fund-raising Lifeboat Day, which was an annual, national event. These atmospheric documentary films, *Boats That Save Life*, and *The Craftsmanship of The Lifeboat Service* show some of these skills in action, with as many as a dozen shipwrights working simultaneously on one boat.

The method of boat construction known as double diagonal involves fixing two layers of planking diagonally over a skeletal frame. White's shipyard did not invent the system but the company did develop its own patented improvements. In the case of these lifeboats the planking was mahogany, which is more suitable for double diagonal than carvel construction for building hulls with compound shapes, such as a double curvature, as it gives greater strength and rigidity needed for heavyweight slipway-launched boats. Mahogany shrinks less than other woods and it is quite flexible, able to be bent, or cold pressed, rather than steamed to shape. In seawater it is fairly resistant to rot, and it will absorb paint, stain or oil. The planks were sawn by machine from logs of mahogany imported from British Honduras (now Belize) which had been that country's principal export through colonial times. Fixed to a 2.5

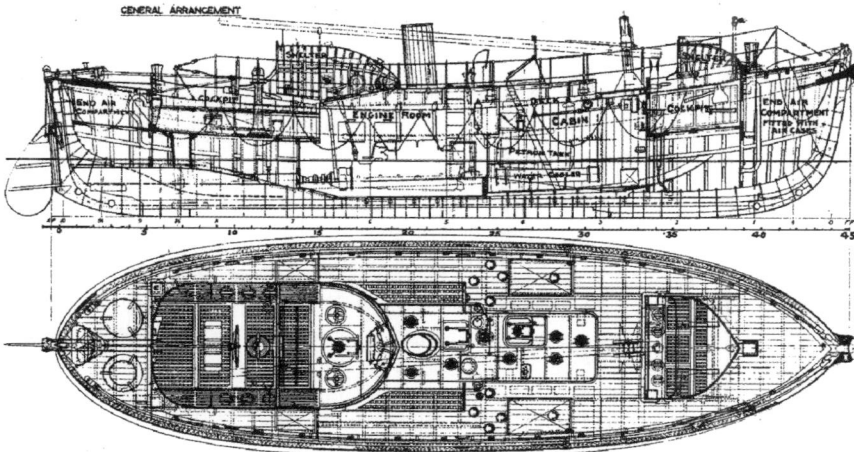

General Arrangement of the 45ft 6inch Watson Cabin class boat

31

ton cast iron ballast keel, the keel of a Watson cabin class lifeboat was made of teak from Burma (another British colony), while the futtocks and ribs were made with Canadian rock elm. The drop keel of the single engined boats had been deleted from the twin-engined vessels. The stem and stern posts were made of English oak, seasoned for three years and shaped by hand, using adzes. The curved 'crooks' for these posts were cut from specially selected branches of oak trees with a natural curve. The deck was also oak.

At the start of construction, the basic frame would be clamped to a wooden 'former' (or template) that sets the dimensions. The first layer of boards, an inner skin known as palings, are either steamed or cold pressed and bent around the completed frame at a 45 degree angle from the keel line, with the top of the plank leading aft. The boards are fixed to the ribs with brass nails. Between that and the outer skin, or 'case', is a layer of unbleached oiled calico (canvas), stretched and glued to provide extra waterproofing. The 'wrapper boards' of the outer skin are then laid in the opposite diagonal from the palings, with the top of the plank leading forward. Usually, both layers of planking would be the same ½" thickness, although some Watson boats used 3/8" thick planks on the inner skin, and ½" on the wrapper boards.

Within the hull there are large air compartments fore and aft, and the design incorporates transverse watertight bulkheads between those compartments and the engine room and cabin spaces. Once the hull is completed, and before any equipment is installed, all the wells and compartments are flooded with thousands of gallons of water to make sure they are watertight. Fitted beneath the side decking and in the

Like other Watson designs, ON736 had twin propeller shafts fitted in shaped cavities known as prop tunnels

bilges were 142 air cases, made of soft wood, covered in calico, painted, varnished and numbered, with spacing bars keeping them apart. These gave added buoyancy and were intended to keep the boat afloat even if the hull was holed, upturned or overwhelmed with sea water. Beneath the waterline there were bilge keels, 20ft long, which add rigidity to a hull and help to keep the vessel upright when it takes the ground, and handrails to be grabbed on to by anyone unfortunate enough to be in the water. Like other Watson designs, ON736 had its twin propellers fitted in shaped cavities known as prop tunnels, which give improved power delivery in shallow water and help keep the boat straight when manoeuvring, as well as providing some protection to the props from floating debris and waterborne objects such as survivors' legs.

The flush deck, without the end boxes fitted to earlier models, was laid with oak planks and there were 10 scuppers fitted along its length to quickly drain water when the decks are awash. The minimal superstructure consisted of two shelters with curved whaleback covers made of diagonal timber and highly varnished. The forward shelter provided space for the bowman and several crew members, while the aft *The yard number,* cockpit canopy covered the instrument panel board, engine controls and *branded into the* engine room hatch, while providing seating for the two mechanics. The *stem post, was* coxswain and assistant cox'n would stand outside in the open cockpit *revealed when the* with a small windscreen for protection from spray; a later adaptation *hull was stripped* was a backrest to lean against and prevent the helmsman being swept *back during* backwards and over the stern. Midships was the funnel, containing *restoration* engine exhaust pipes, and three large ventilator cowls, and forward of

W
1705

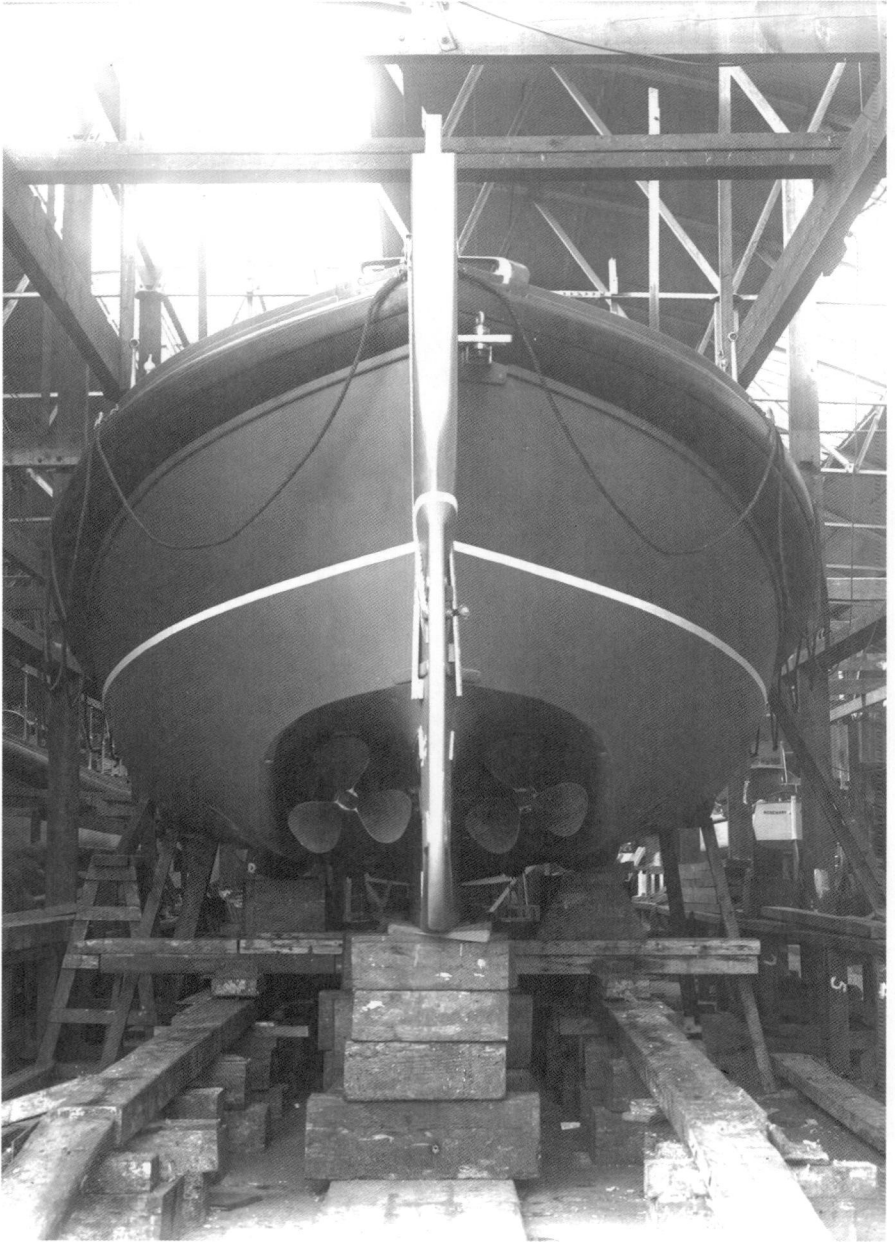

Before and after: A Watson boat nearing completion and 'our' Watson being resurrected

that was a hatch cover to the companionway leading down into the cabin. Farther forward was a locker for the anchor chain, and the tabernacle or step for the foremast. The cork belting, or 'wale' that ran around the hull to act as a fender, was rounded cork, measuring 13in x 6in. The rudder was hung externally and extended beyond the 45ft 6in length overall.

As this class of boat was designed for slipway launching, the exposed rudder was vertically adjustable on the rudder post, to be lifted out of harm's way and dropped into the water by the assistant (second) mechanic on launch and recovery. There was also no need for anti-foul paint, as the boat would be washed down after each service and kept out

of the sea water, so the hull bottom would be painted gloss white.

When she was commissioned, the *W&S* was supplied with a staysail and topsail, along with four oars made of round ash. This was a throwback to the Watson's sailing and pulling origins and there is no record of any of these being used during our boat's working life. The oars were lashed to the port side railings, while on the starboard side the sea anchor was kept. On later models the sails were dispensed with altogether, taking the boats truly into the 'modern' age.

Before acceptance trials the completed boat would be subjected to stability tests, when heavy weights are fitted to the gunwales, one side

A Watson Cabin class boat taking shape at White's workshop,
RNLI PHOTO (left)

at a time, to see how far the boat will lean under its full load. On a small commercial vessel these tests will determine the number of passengers and total cargo permitted on board; with a lifeboat this hardly applies, as the boat will take as many people as it finds in need of rescue.

The finer points of engine design

The Watson boats themselves were the result of a century's evolution, but the engines that powered the cabin type boats had been developed over a short period in the 1920s. The first Watson motor lifeboats, the 38ft, 40ft and 43ft types, were based on the established pulling and sailing hulls, with the simple addition of an engine and propeller as an auxiliary. Engines were supplied by Tylor, Blake and Wolseley, and the 40hp Tylor became the preferred choice, giving a speed of around seven knots. Tylor engines even replaced Blake engines in two boats in 1914.

In 1922 the RNLI discussed the important difference between the designing of a lifeboat and the designing of her engine. The *Life-boat* Journal reported: 'The requirements of the Institution, so far as hulls are concerned, are quite special. No other small craft are built which would be in any way suitable for the Lifeboat Service. It is very unfortunate that there does not exist any standardised engine on the market which meets the requirements of small open boats and is, therefore, suitable for Lifeboat work. The result is that it was considered best for the Institution to design one for itself. A design was made, and specifications were drawn

out, at the beginning of 1920, and Messrs. J. Tylor & Sons, Ltd., who had made practically all the previous Lifeboat engines, were instructed to build the engine. Unfortunately, they had to discontinue the work and the Institution took over to complete the work itself.'

Detailed drawings of the various parts were issued and contracts placed for their supply. 'When completed, they were delivered to the

The preserved H.F. Bailey, now retired at Cromer Lifeboat Museum, made an unscheduled stop at Penlee in 1957

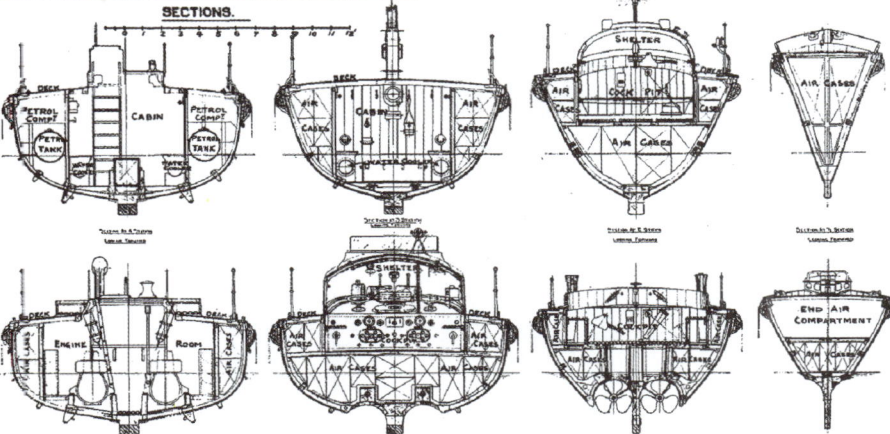

45ᶠᵗ 6ᴵᴺ x 12ᶠᵗ 6ᴵᴺ TWIN SCREW CABIN MOTOR LIFEBOAT.

SECTIONS.

Weyburn Engineering Co, Ltd, of Godalming, Surrey to which was entrusted the work of erecting and testing the engine, which was rather an anxious business. We were very fortunate in finding nothing more serious than some faults in the starting motor, which was bought from an outside firm, some faulty piston rings, and carburettors which required rather a long time for adjustment. Unfortunately, while the engine was receiving its final coat of enamel, before being sent to Cowes to be installed in a lifeboat, part of the Weyburn works were burnt down, and the engine was seriously damaged.'

The features which the RNLI considered essential were as follows: 'The engine is spray-proof, all working parts, ignition details, and other items which can be effected by damp, rust or dirt, being enclosed in a water-tight cover, and if there are any actual moving parts, they are supplied with a copious bath of lubricant, the lubricant being pumped up from the bottom of the crank chamber by a very powerful pump, and, after being filtered and cooled, distributed throughout the whole of the engine.

'The only moving part that can be seen or touched is the flange on the thrust shaft at the after end of the engine, and this passes into the reverse gear through a substantial gland. For this reason the engine is submersible, the limit of submersibility being the air intake to the carburettor. It is dangerous to have hot pipes or parts on a marine engine, besides being uncomfortable and unsatisfactory, and, therefore, every hot part is enclosed in a water jacket. A simple apparatus has been added to this engine which automatically keeps the cooling water in the jackets at the same temperature by regulating the amount which it allows to pass through.

'The engine ought to have its reverse gear as an integral part of the engine to keep it absolutely in line, and it is necessary to provide a separate oiling system. The thrust block is a very vulnerable point in the usual marine engine, and this block also should be part of the one system, so that it can share in the adequate oiling arrangements provided for the engine.

'In a marine engine, and especially in a Lifeboat engine, the control should be as obvious and as easy to operate as are the controls in a motor car, and for this reason inter-connection has been arranged between the throttle and control gear so as to leave one operation only for the driver to attend to. This involves a certain amount of mechanical work which may appear to have complicated the engine, but it has simplified the handling of it.'

Despite this intention, the controls on the Watson Cabin boats could not be operated by the cox'n alone. Each of the twin engines needed a mechanic to operate it. The Motor Mechanic would work the starboard controls and his assistant took the port side. They worked deep inside the cockpit canopy, operating duplicated controls each side of the engine room hatch, under the command of the cox'n. The aft cockpit was exposed to the elements and in heavy seas would have been awash with sea water and spray. The howl of the wind, the throb of the engines, rotating prop shafts and roar of the exhaust would have meant the helmsman needed

A four-cylinder Weyburn engine, preserved at the Chatham Lifeboat Collection, looking almost as pristine as the ones to which Johnny Drew devoted so much of his time

PHOTO: IAN SMITH

to shout out instructions to the men crouched beneath his knees, while trying to listen out for warnings from the bowman, who would have been almost invisible, covered in spray some 40 feet ahead of the cockpit.

The controls on each side contained a start switch, stop-start control, wheel-operated reverse gear control, maximum and minimum throttle control, electric fuel pump switch, gauges and instruments with board light, and with main throttle levers at the bottom of the control board. Midships and aft of the control board was the steering wheel and binnacle, and above and around the engine room hatch was the separate light switch, hooter switch, clock and fire extinguisher. To the port was the wave subduing oil pump handle, the voice pipe from the cabin, and to starboard was the coxswain's signal gong to the mechanic and a loudspeaker from the cabin.

Starboard prop tunnel on the W&S during renovation, and the finished article on Thomas McCunn, preserved at Longhope

The RNLI report continued: 'Great trouble has been experienced in the past in starting some of our larger engines on a cold morning. The new engine, therefore, has been fitted with a starting gear which is not dependent on outside batteries or bottles of compressed air, but which consists of an ordinary bicycle motor.

'While no adjustment should be made to a lifeboat engine while actually running on service, every effort has been made to allow parts which require adjustment, examination, or replacement, put in such a position that they are accessible. For instance, the whole of the pistons can be removed from the engine through the crank chamber doors, and this has actually been done in one hour.

'In order to provide for a thorough overhaul of the engine without removing it from the boat, the design includes a detachable cylinder head, which contains all the delicate mechanism, such as the valves, ignitors, coils, etc, while all pipe-joints for the exhaust, water and oil circulations are made on the engine body. This allows the head to be removed, the valves ground in, and adjustments made without breaking any joint other than the main cylinder head-joint which is a plain soft copper grummet, as it is called, and the head can be replaced without upsetting the adjustment of the parts.

'In 1925 the RNLI decided that all the larger types of Motor Lifeboats should be built with two engines each, and this made necessary a lighter engine. Two engines, or rather two variants of the one engine, have now been designed. The one has six cylinders giving 60 h.p. and the other four cylinders giving 40h.p. While all the larger types of Motor Lifeboat — those over 40 feet —- will for the future have either two 60 h.p.

Looking forward from the funnel, ventilators, capstan and hatch to the survivors' cabin

COURTESY LONGHOPE
LIFEBOAT MUSEUM

From racing car to rescue boat

A later article in *Motor Sport* magazine explained the genesis of this engine. In 1925 Major F. B. Halford, who became Chief Engine Designer at de Havilland, developed a 6-cylinder engine for his own racing car. It featured a 7-bearing crankshaft and H-section forged duralumin con-rods with white metal big-ends, and two inclined overhead valves per cylinder were operated by twin o.h. camshafts. Ignition was by two magnetos, mounted vertically, one driven from the rear of each camshaft, firing special 12-mm. spark plugs, two per cylinder. This engine was built by The Weyburn Engineering Co., already known for lifeboat and aero engines and making camshafts for car manufacturers. The Halford Special was fitted in an Aston Martin chassis with racing bodywork and it competed at Brooklands with some success.

A much less exotic version, the Type AE6, with cast-iron crankcase and overhead valves, was developed for the RNLI, and two prototypes were assembled at Elstead and tested in 1929. The following year the first production engine was installed by Saunders Roe in a 35 ft. 6 in. Liverpool class lifeboat (*City of Nottingham* ON726), in a watertight engine room beneath pent roof access hatches ahead of an aft cockpit shelter from which the mechanic operated the engine controls. About 20 of those 35hp, 6-cylinder engines were fitted over the next three years, but almost simultaneously Weyburn had put into production the CE4, a 4-cylinder version, two of which were being installed in Yard no 1705 at J.S. White's over the winter of 1930/1931. The smaller and lighter CE4 gave similar performance to the six-cylinder motor with better economy.

The prototype 6-cylinder engine that was damaged by fire was subsequently completed and fitted to the Penlee boat *The Brothers*, which made its first service in January 1923, saving 27 seafarers from the Yugoslav steamer *Dubrovka*, wrecked near Land's End. Among the lifeboat's crew, on his maiden shout, was Johnny Drew, who went on to become totally engrossed with the Weyburn engines in his beloved boat, the W&S.

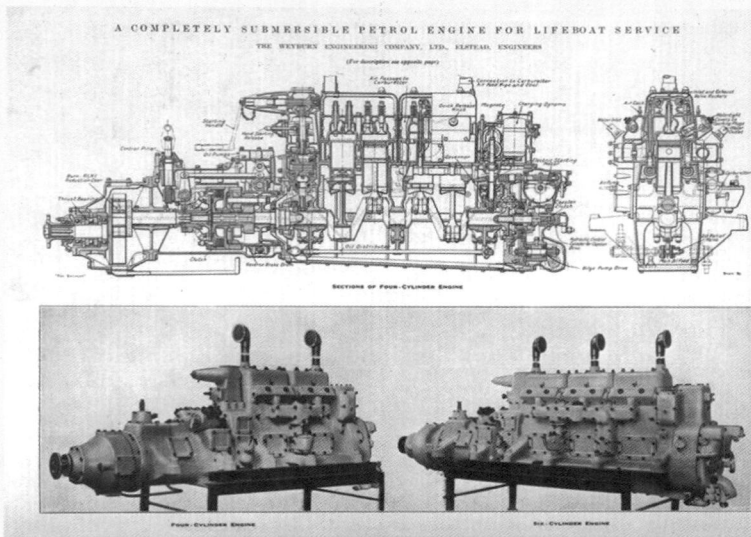

A COMPLETELY SUBMERSIBLE PETROL ENGINE FOR LIFEBOAT SERVICE

THE WEYBURN ENGINEERING COMPANY, LTD., ELSTEAD, ENGINEERS

SECTIONS OF FOUR-CYLINDER ENGINE

FOUR-CYLINDER ENGINE

SIX-CYLINDER ENGINE

or two 40 h.p. engines, the smaller will have one or other of these two engines, except the 35-ft Motor Life-boat, which, can be launched from a carriage. For this specially light type a third and lighter 35 h.p. Halford engine, adapted to the Institution's requirements, is to be used.'

The *Life-Boat* Journal records that the first of the new 4-cylinder engines, two of 40 hp, were installed in the new Watson Cabin lifeboat *K.E.C.F.* (ON700) for Rosslare Harbour in Ireland, and this new boat and her engines were submitted to a very severe test. The engine-room was flooded with fresh water up to the outside water-level, the water weighing 5 tons, the equivalent of 83 men. In this condition, and with 17 men on board, so that the total weight represented 100 men, the lifeboat was taken to sea — a moderate wind was blowing and the sea was slight — and for two hours she ran at her full speed of 1200 rpm. The 5 tons of water in the engine-room was washing heavily over and round the motors and against the bulkheads, but every one on board felt complete confidence in the boat's stability and sea-worthiness. At the end of the trial the water was pumped out, and the next day the lifeboat was taken out for her endurance trials. A southerly gale was blowing, with a very heavy sea. She was tried with the sea ahead, on the bow, abeam and astern, and although nothing whatever had been done to her engines since the flooding, she ran for 8 hours at full speed, under these varying conditions, without a hitch.

Following acceptance trials the *W&S* was sent to the Penlee station on February 26, 1931. Her story starts there.

The W&S on acceptance trials on February 25, 1931, with inspector on board, photographed by Beken of Cowes
RNLI PHOTO

Name *R.N.L.B. W. and S.*

LIFE BOAT No. 736

Registration Nº 162590.

1.—Donor. *Legacies of Miss W. A. Coode & Miss Ellen Young*

2.—Cost. *£7684-7-2*

Builder's Initial and No. *151705*

3.—Builder. *J. S. White & Co. Ltd*

4.—When taken over. *26th February 1931*

5.—Type of Boat. *Watson, Cabins, Motor*

Beam extreme to outside of planking. *12' - 6¼*

6.—Length over all. *45' - 5½"*

To top of end boxes at Stem Head and Stern Post. *F. 9'-1". A. 8'-3¼*

7.—Depth:—Amidships, bottom of Keel to Gunwale Capping. *6'-4"* F. 2'-11⅛ F. 9'-2¼ Do. Below F. 10½ horizontal line A. 4⅝

F. 5'-5¾ Width at Gunwale. A. 9'-6"

8.—End Boxes:—Length from inside Posts. A. *5'-3¼* Height of Bulkhead above Deck. A. 2'-3¼

9.—Height:—Lower Side of Skin *5'-6⅞* Upper side of Deck to upper side of Thwarts. *Not so fitted* Upper side of Thwarts to Gunwale Capping. *Not so fitted* to upper side of Deck at side

10.—Relieving Tubes:—Number. *Scuppers & Tubes - 10* Size. Width at Top and Bottom. *12⅜* Weight of Iron. *T. c. q.* *2. 5. 3. 0* Length of straight part of bottom *32*

11.—Keel:—Depth. *9⅞* Wood. *4¼"* Iron. *5⅞"* Length. Drop at after end. Weight. Material.

12.—Drop Keels. *None fitted*

Fore Drop Keel.

After Drop Keel.

13.—Bilge Keels:—Spread from centre to centre at bottom. *6' 9"* Length. *20'* Depth. *6⅛* Width at top and bottom. *3¹/₁₆* Height of bottom at centre from bottom of keel *9¾*

14.—Thwarts:—Spacing from centre to centre. *Not so fitted* Head and stern sheets to centre of thwarts.

15.—Water Ballast:—Description and amount. *- do -* Number of Tanks. Length. Breadth. Depth.

16.—Wale:—Description and Dimensions. *Solid Cork 13" × 6"*

17.—Weight of Boat complete, without gear.

18.—Rig of Boat. *Staysail & Trysail*

19.—Oars:—Number and Description. *4 Round Ash*

20.—Date and place of Harbour Trial.

21.—Draught of Water:—Light. *32 30½* Fore. Aft. Down tubes. Crew and gear in and tanks empty. Fore. Aft. Down tubes. Crew and gear in and tanks full. Fore. Aft. Down tubes.

22.—Tests answered:—

Gross Tonnage 17.78. STABILITY.

Register - - - 9.36.

Tanks full. Tanks empty.

Number of Men on gunwale to bring it awash, with crew and gear in place ...

Ditto to bring deck ditto ditto ...

SELF-RIGHTING POWER.

Tanks full. Tanks empty.

Number of Crew in place with all gear and sails stowed ...

Ditto with sails set, jib and mizen sheets fast, boat hove over degrees

Number of men on lee gunwale in addition to crew in place, sails set, &c. ...

DRAUGHT OF WATER—BOTTOM UP.

Forward ... ... Tanks full. Tanks empty.

Aft ... ...

23.—Name of Station. **PENLEE**

24.—Date sent to Station. *26th February 1931* Date of Test Exercise.

25.—Date and cause of removal from List.

Result.

How disposed of.

LIFE BOAT No 736   Name RNLB *W and S*

Registration No. 162590

1.-Donor. Legacy of Miss A Coode and Miss Ellen Young
2.-Cost. £7,684-7-2
3.-Builder. J.S.White &Co Ltd   Builders initial and No W1705
4.-When taken over 26th February 1931
5.-Type of Boat Watson, Cabin, Motor
6.-Length overall 45'-5½" Beam extreme to outside of planking. 12'-6 ¼"
7.-Depth:- Amidships, bottom of keel to Gunwale Capping. 6'-4"
          To top end boxes at Stem Head and Stern Post. F 9'-1" / A 8'-3¼"
8.-End Boxes;- Length from inside Posts F.5'-83/4" / A.5'-37/8"
          Height of Bulkhead above Deck. F. 2'-113/8" / A.2'-3¼"
     Width at Gunwale. F. 9'-2½"A.9'-6" Do. Below horizontal line. F.103/4" /
     A.61/8"
9.-Height: - Lower side of skin to upper side of deck 5'-61/8"
          Upper side of Deck to upper side of Thwarts 'Not So Fitted'
          Upper side of Thwarts to Gunwale Capping 'Not So Fitted'
10.- Relieving Tubes:- Number. Scuppers & Tubes – 10
11.- Keel:-Depth 9 7/8". Wood. 4¼". Iron. 5 5/8". Width at Top and Bottom 12 1/8"
          Weight of Iron. 2T. 5.3.0. Length of straight part of bottom 32ft
12.- Drop Keels None fitted
13.- Bilge Keels: Spread from centre to centre at bottom 6'.9 1/8". Length. 20'. Depth.
     6 1/8" Width at top and bottom. 3 1/16" Height of bottom at centre from
     bottom of keel. 9¼"
14.- Thwarts:- 'Not so Fitted'
15.- Water Ballast 'Not so Fitted'
16.- Wale:- Description and Dimensions. Solid Cork 13" x 6"
17.- Weight of Boat complete, without gear. Gross Tonnage 17.78, Register 9.36
18.- Rig of Boat. Staysail and Trysail
19.- Oars:- Number and Description. 4 Round Ash
20.- Date and place of Harbour Trial.
21.- Draught of Water:-Light. Fore 32". Aft 50¼"
          Crew and gear in and tanks empty. Crew and gear in and tanks full Fore Aft .
22.-Tests answered:-
Gross tonnage   17.78
Register   9.36

STABILITY N/A
SELF RIGHTING POWER N/A
DRAUGHT OF WATER N/A

Name of station PENLEE
Date sent to station 26th February 1931

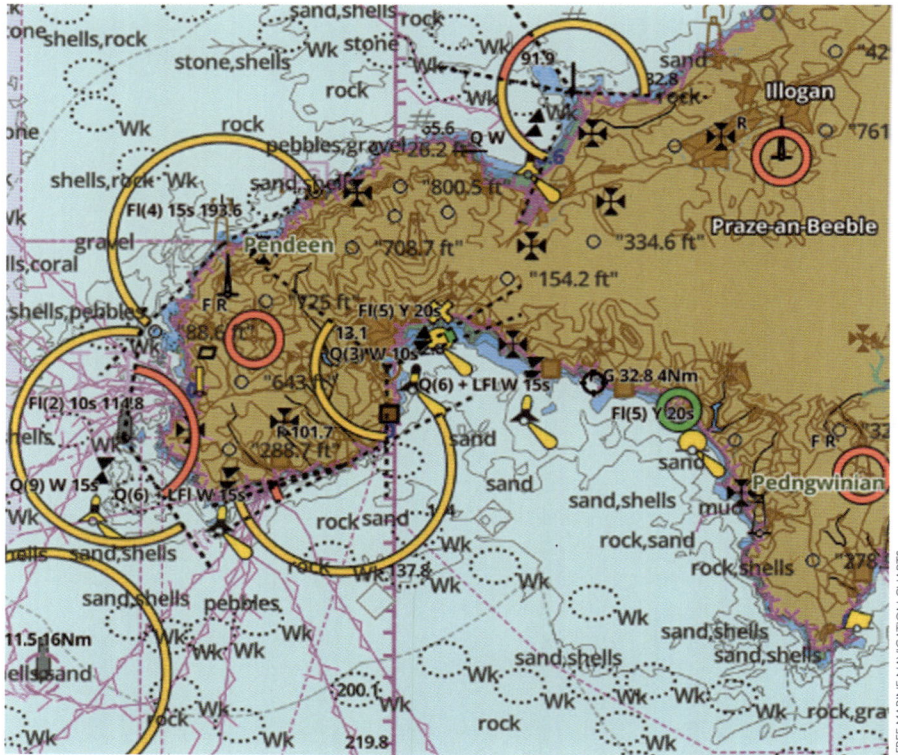

The Penlee Patch

The Boat's working waters

THE territory covered by the Penlee lifeboat, known as its 'patch', includes the total area of Mount's Bay and extends southwards into the English Channel and westwards towards the Isles of Scilly. Mount's Bay is the largest bay in Cornwall and its west side is sheltered from the prevailing Atlantic Westerlies, leaving the opposite east coast exposed to Southerly and South-Westerly gales. This rocky coastline is a dangerous lee shore which was the graveyard of many a ship during the days of sail. The coastline extends about 42 miles (68km) from Lizard Point: position 49.96N, 5.21W, to Gwennap Head 50.03N, 5.68W.

The Wolf Rock lighthouse. Before it was automated in 1988, keepers were transferred by Breeches buoy from TH launches

The principal towns in the Bay are Penzance, including its neighbouring villages of Newlyn, Mousehole and Paul, to the west, and the more central Marazion, facing the island of St Michael's Mount, from which the Bay takes its name. To the south and west of Mousehole, towards Land's End, the granite cliffs contain the villages of Lamorna and Porthcurno, and adjoining coves, which in the mid-20th century claimed their own quota of lost vessels. To the eastern side of the Bay lies the notorious three-mile long Loe (Looe) Bar, fronting Cornwall's largest fresh water lake at Loe Pool. Over the years the beach there has claimed more than 100 wrecks and many hundreds of lives, most famously the wreck of the *Anson*, that inspired the development of the Life Saving Apparatus, which established stations at The Lizard, Porthleven and Newlyn. In the days of sail, an extreme Southerly gale could force eastbound ships

into this part of the Bay, where they would risk becoming 'embayed', with their only hope of escape to turn about and spend days trying to tack back towards Penzance in the sheltered western corner of the Bay.

Mount's Bay has an intriguing pre-history: On either side of Penzance, on the beaches at Ponsandane and Wherrytown, fossilised tree trunks from a 'submerged forest' can be seen at low tide. Farther out, Gwavas Lake is an area of relatively calm water resulting from a fresh water lake existing in the ice-age. More recently, this area of Cornwall endured invasion during the Anglo-Spanish war (1585-1604) when Spanish forces sacked and burned Newlyn, Mousehole, Penzance and Paul. In 1625 'Turks', or Barbary Pirates from North Africa, kidnapped about 60 men, women and children to be sold as slaves to the Ottoman Empire. Twenty years later another 240 locals were reportedly captured. In total, the Barbary Pirates are believed to have seized more than a million Europeans, with Cornwall targeted as easy pickings, both afloat and ashore. The 'Turks' attacked so frequently and brazenly in the 17th Century that some fishermen refused to put to sea. Eventually, peace was brokered with Tripoli in 1675 but it was not until 1816 that the Barbary Wars finally ended that slave trade and repatriated some 4,000 European Christians.

In November 1755, the Lisbon earthquake created a tsunami that caused the sea-level in Mount's Bay to rise by 10 feet (3m) at great speed and it ebbed at the same rate. In more normal times the coast of west Cornwall is prone to sudden, but less dramatic, changes of weather and a variety of micro-climates, producing sudden squalls, patches of fog or sea mist, often accompanied by heavy ground swell rolling in from the Atlantic. One source has estimated that as many as 30 severe storms can be expected during a bad winter. Following a South West gale these conditions have been described as 'Wreck weather'.

To the west of the mainland, the Isles of Scilly were often the first land sighted by vessels arriving at the Western Approaches from the Atlantic, and very often those rocky outcrops were also their final landfall. Those notorious waters proved to be the graveyard for hundreds of ships over the years. One of the most infamous episodes being the loss of Sir Cloudesley Shovel's squadron of the British Fleet in 1707 in which 2,000 men died. As an aid to navigation, Trinity House, the organisation which administers the lighthouses around England (founded by Henry VIII in 1514), eventually decided to build a lighthouse on the most westerly island of Bishop Rock to augment the only existing light at St Agnes. The first construction built on cast iron piles proved inadequate and the lighthouse was subsequently built with granite blocks on a design based on the Eddystone lighthouse and opened in 1858. Offshore hazards are many in this area and only a few are marked by shore-based lighthouses on headlands, along with the only remaining light vessel, the Sevenstones.

Since the 14th century there had been a beacon burning from the chapel at the high point of Carn Brea, and another at St Michael's Mount, but with King Henry's dissolution of the monastries, church beacons were no longer permitted. Those lighthouses covering the far south west of Cornwall are: Longships, on rocks 1.25 miles West of Land's End (built in 1795); Wolf Rock, 8 miles SSE of Land's End (1862); Tater Du, on the shore to the east of Land's End (1965); The Lizard on shore (1619 and 1752). The Sevenstones lightship (LV19) is one of the last of these unmanned vessels remaining in British coastal waters, positioned due West of Land's End at 50 03. 616N, 6 03.337W.

The two biggest hazards to shipping in our boat's patch are the Wolf Rock and the Runnelstone. The Wolf Rock lighthouse was designed by James Walker and completed in 1869 by the engineer/constructor James Douglass, who also re-built the quays at Mousehole harbour. The name is said to be derived from the howling sound effect created by wind and waves entering voids in the rock structure. Trinity House established a depot at Penzance at that time, primarily to supply the rock used in the tower's construction, and this base remained functioning until it became a museum and closed in 2004. Post WW2 the Wolf Rock became the world's first lighthouse to be equipped with a helideck above the lantern. It features in several escapades that involved 'our' lifeboat in the post WW2 years and has for long been a meeting point for medevac rendezvous. It was automated and de-manned in 1988.

The Runnelstone, previously known as the 'Rundle Stone', is a granite pinnacle one mile south of Tol-Pedn. The name is thought to be taken from the Rennell Current (after the scientist James Rennell), a tidal stream which runs up from the Bay of Biscay and divides into two branches south of the Scillies, one stream flowing into the English Channel and the other to the south of Ireland, and which can push vessels several miles to the north of their intended course. Another theory is that the name comes from the Cornish *Men Reunel*, 'stone abounding in seals'. In the days of sail it was almost an annual event for some unfortunate vessel to be lost on the Runnelstone and the stone's appetite for wreck continued into the 20th century as steam predominated, with some 30 steamers coming to grief here in 40 years, culminating in 1923 when the 6,000 ton Ellerman Line's *City of Westminster* knocked the exposed top off the stone. That wreck was attended by the Penlee lifeboat *The Brothers* (predecessor to the *W&S*), skippered by Frank Blewett. The stone was a hazard facing vessels travelling in any direction, which now lies 20ft below the surface, and is marked by a buoy and daymarks on shore at Gwennap Head. Belatedly, a lighthouse was eventually built on shore at Tater Dhu in 1965, to indicate the presence of the stone.

The Bay of Wrecks

The many dramas that occurred offshore around those obstacles generally involved ocean going vessels, but within the Bay itself there have been many and various incidents befalling vessels of all description. There have been estimates that some 300 coastal vessels and fishing boats have been wrecked within this area over the years but any figure would be pure guesswork. In 1899, on Thursday April 7, a Force 10 NW gale raged all through the night, and early in the morning of the 8th a clutch of fishing boats moored in Newlyn and Penzance harbours went adrift at about 4am causing 'much activity in rousing crews and fishermen to secure their boats'. A total of 10 luggers were blown off their moorings and ended up scattered along the coast from The Lizard to Lamorna. They were all PZ registered fishing boats: the *Come On, Cygnet, Dart, Dewdrop, Excel, Florence Edith, Onward, Queen of The Day, Sir Wilfred Lawson* and *Valleta*, all wrecked within a couple of hours.

Most of the ill-fated vessels that have been wrecked in our lifeboat's patch either foundered in deep water, where they eventually broke up or sank into the sand, or they were driven ashore to be broken by natural or man-made forces. Some, including the WW2 casualty HMT *Royalo*, were considered hazards to navigation and subsequently blown up. Few wrecks have survived in any recognisable form, but in 2006-2008 a survey conducted by the Cornwall and Isles of Scilly Marine

Archaeological Society (CISMAS) investigated the inshore waters of the Bay and made some dramatic sidescan sonar images of the steamship *Hellopes*, which came to grief right outside Mousehole harbour and had remained largely intact for almost 100 years.

    This ship had previously been damaged when grounded off the Cape of Good Hope and was making its final voyage towards the breakers at Falmouth in December 1911, when its cargo of coal shifted, generating an irreversible list which caused it to founder. The crew of four were rescued by the *Elizabeth and Blanche II*, only two weeks after the lifeboat had gone to the assistance of the Norwegian barque *Saluto*, wrecked under the cliffs near Penlee Point.

Holds   Engine   Boilers   Hull plates   Mast   Bow

History of Penzance-Penlee branch

The first lifeboat in Cornwall, partly funded by the insurance company Lloyd's of London, was delivered to Penzance in 1803. It was a 27ft long, Greathead-designed North Country class pulling and sailing boat, which was kept near the eventual site of the Penzance railway station. But after languishing unused for nine years the boat was seized for debt and sold. It was not replaced until 1826. Whatever service was previously available to vessels in distress would have been provided by the local 'hovellers' — a term applied in Cornwall to freelance boatmen who plied for hire as unlicensed pilots, 'seeking' vessels in need of assistance, and who may not be 'true fishermen'. Another definition of the term suggests that some of those boatmen might have dabbled in less savoury, or legal, pursuits. The *Sailor's Word-Book* says the term can be 'applied colloquially to sturdy vagrants who infest the sea-coast in bad weather in expectation of wreck or plunder'.

The Lloyd's agent Richard Pearce also carried life-saving gear on his own gigs, which he would send out on rescue missions. In 1824 a branch of the National Institution for the Preservation of Life from Shipwreck (predecessor of the RNLI) was formed and a new boathouse was built nearby, which housed a Plenty class lifeboat from 1826 to 1828, when it was wrecked on launching. From then on, there was another long period without a boat until Richard Pierce, now the branch Hon. Sec., raised funds for a 10-oared Peake class vessel that served from 1853 to 1860, moving in 1856 to a new boathouse that was built by the RNLI for £88. From 1860 to 1865 the first named boat to be based here was the *Alexandra,* which rescued eight seamen from the *Willie Ridley* of Plymouth, earning Capt. T. H. Fellowes a Silver Medal. After difficulties launching to the aid of several vessels in Mount's Bay, the station was again moved along the sea front to Wherrytown in around 1862.

Over the next 20 years the 10-oared, 24ft boat *Richard Lewis* saved 86 lives. In 1866, one of the most dramatic lifeboat shouts in Britain saw the *Richard Lewis* being hauled by a team of eight horses nine miles overland to Hayle on the north coast, where she assisted the St Ives lifeboat in saving five lives from the SS *Bessie,* which earned another Silver Medal; and more medals were to follow. In 1884 the RNLI decided to relocate yet again to a new granite-built boathouse erected close to Penzance harbour, at a cost of £575–6s–6d. This was paid for by a £1,000 gift from Henry Martin Harvey, which also paid for a new lifeboat, the *Dora,* given the Official Number 49, and a carriage.

The boathouse built on the waterfront at Penzance in 1884 survives, albeit with a different function from its original purpose

The *Dora* served from 1885 till 1892, saving 57 lives. *Elizabeth and Blanche* (ON378),

was the last self-righting boat to serve here from 1895 to 1899. The subsequent vessel, *Elizabeth & Blanche II* (ON424) was a 38ft Watson class, also pulling and sailing type but with deeper draught. It carried 12 oars (six per side) and weighed five tons. On November 1, 1907 the Thames sailing barge *Baltic* went aground on St. Clement's Isle just outside Mousehole harbour. Attempting to launch to its rescue, the lifeboat became stuck on its carriage on the muddy beach, despite having 10 horses and countless helpers to manhandle it. Some Mousehole men organised a rescue with the crabber *Lady White* and all hands were saved. As a result of that failure to launch, in 1908 the Penzance Committee decided to move the lifeboat to Newlyn, where it was kept on its carriage under a tarpaulin beside the harbour.

On Boxing Day, 1912, the SS *Tripolitania*, went ashore on Loe Bar, near to Porthleven. The Newlyn boat *Elizabeth and Blanche II* could not get out so Cox'n Will Nicholls launched the *Janet Hoyle* (ON386) from Penzance to stand by. A 90mph gale was raging and Penzance

On Boxing Day 1912 the Tripolitania was wrecked on the notorious Loe Bar but the weather prevented the Elizabeth and Blanche II from launching, so the Janet Hoyle was launched from Penzance

The Elizabeth and Blanche II was the last pulling and sailing lifeboat to be stationed in Mount's Bay and the first to occupy the new boathouse at Penlee

The first lifeboat in Cornwall was a Greathead designed pulling and sailing boat

Pier Head was under water. Two of the boat's crew died a few days later from pneumonia, which shows the terrible conditions they had to face on that service. The *Elizabeth & Blanche II* saved a total of 72 lives between 1908 and 1912 from Newlyn, before moving to yet another new location. The 1884 boathouse remained as a reserve station until 1917, housing the *Cape of Good Hope* (ON341) which saved nine from the steam drifter *Renown* in 1909, and later the *Janet Hoyle* from 1912-1917.

The *Elizabeth and Blanche II* remained at Newlyn until 1913, when a new boathouse was opened at Penlee Point, two miles south of Newlyn on the road to Mousehole. The RNLI had surveyed the Penzance area in 1910 and found that this cliff-side location provided the best opportunity to launch a boat from a slipway under any state of tide or sea condition. The roadside stone-built building faces east across Mount's Bay, and its slipway is believed to have been the second steepest in Britain.

The boat was rowed to the new location by the Newlyn crew, who then reluctantly handed the vessel over to the men of Mousehole. The Newlyn men, however, were first given the chance of a ride down the slipway, which saw them drenched as the boat plunged into the water from its full height at great speed before disappearing beneath the waves. First a hat floated to the surface before the boat itself rose out of the foam, carrying the disgruntled party, to the amusement of those ashore. Embarrassingly, it proved difficult to rehouse the boat and it returned to a mooring at Newlyn.

The first shout for *Elizabeth and Blanche II* at Penlee was not until October 1914, when she went to the assistance of the SS *Liguria* which had lost a propeller. This lifeboat proved to be a popular craft and when it was retired in 1922 there were plans to convert it to a long distance cruising boat, and there were reports in 1925 of it being prepared by Captain Nicholls for a world cruise, but there is no record of that ever happening. All together 23 lives were saved from Penlee before the old pulling and sailing boat was replaced by the first motor lifeboat in Cornwall, *The Brothers* (ON671).

A model of a Henry Greathead's 'Original', displayed at the ceremony marking Penlee's 150th anniversary. Delivered in 1803, it was the first lifeboat to serve in Cornwall, and the first boat built specifically for saving life. It was a 30ft self-righting design, pulling and sailing boat with built-in buoyancy compartments, carrying 10 oars, and was partly funded by Lloyd's of London. It was delivered to Penzance in 1803 from builders in the North East, but its unfamiliar design was not popular with local boatmen and it lay, unused, on the beach at Penzance for nine years.

The *Richard Lewis*, which replaced the *Alexandra* at Wherrytown in 1865 was a 24-ft self-righting boat, crewed by 13 men with 10 oars. Its first shout was in the winter of that year to the brigantine *Tobaco*, saving five lives, and early in the New Year it made the dramatic journey overland to Hayle. In 1867 it saved 30 lives from five shouts and more dramas followed, with its most celebrated rescue being that of the *North Britain* driven ashore between Long Rock and St Michael's Mount. In 14 years it saved 57 lives.

The deeper draught, pulling and sailing type boat *Elizabeth and Blanche II* was the first non-self righting boat to serve at Penzance, which replaced its namesake in 1901. Designed by Watson and built on the Isle of Wight, its design was influenced by Philip Nicholls, one of Watson's yacht skippers and a Penzance pilot. It was moved from Wherrytown to Newlyn in 1908, from where it made 12 services, saving 72 lives before relocating a second time to Penlee. After service it was bought by Captain Nicholls.

The *Brothers* was one of the first generation of motor lifeboat, powered by the Weyburn engine. It was a Watson design, built by J. S. White at Cowes in 1920. It served at Penlee from 1922-31, then at Falmouth from 1931-34. It was put on the Reserve fleet from 1934-48, and finally at Workington from 1948-52. After service it worked as a dive support boat at Dover and in 2020 was brought back to Cornwall to await restoration at Gweek boatyard.

Wreck of the Baltic

Johnny Drew, who was to become Mechanic on the *W&S*, tells how the wreck of the Thames barge *Baltic* on St Clement's Isle in 1907 influenced the RNLI's decision to move the lifeboat to a permanent boathouse. The 55ton barge (built at Battersea in 1893), carrying cement from the Medway to Newlyn for the construction of the harbour wall, went aground in weather so bad their plight went unnoticed until the crew lit a fire to attract attention.

"When the *Baltic* came ashore it was the small crabber *Lady White* that was launched from Mousehole harbour to rescue the crew," recalled Johnny. "The *Elizabeth and Blanche* was on the carriage under the Red Lion at Newlyn. That's where she used to be kept on a carriage, on big wheels. When she was launched the covering would be taken off, the crew would be aboard and she would be pushed down the slope to the harbour. But on this occasion she went down and the big wheels stuck in the mud and they couldn't get her off."

Aware of the need for action, six Mousehole residents — Johnny's father Stanley Drew, Dick Thomas, Luther and Harry Harvey, and Richard and Charles Harry — took the initiative to launch the *Lady White,* a heavy, clinker-built boat, using the harbour crane to drag it over the protective baulks which lay across the harbour mouth and rowing out to rescue the barge's four crew, including the captain (probably George 'Doggy' Fletcher), his wife and daughter, along with the mate Adam Torrie. They were brought ashore to the cheering of local spectators.

Adam Torrie was lodged at the home of Francis Blewett, the local harbour master and father of Frank, who would later become Coxswain of the Penlee lifeboat. The crew of the *Lady White*, who were not RNLI crew members,

were acknowledged by the Institution, with awards of £2 each. And a visitor to Mousehole, Mr Pellow, who witnessed the event, had silver and gold medals struck for each of them. Mr Torrie was evidently made very welcome at the Blewett household as he went on to marry Frank's sister Janie. Their grandson Barry became a crew member of the Penlee boat and Stanley Drew (pictured above, with Johnny to his right) joined the launch team at the new boathouse.

The *Baltic* was refloated and returned to trading on the East Coast, once needing assistance from the Harwich steam lifeboat in December 1910. On the centenary anniversary of the rescue in November 2007, a commemorative stone plaque was unveiled on the Quay at Mousehole.

The Penlee boathouse

After all the problems launching the lifeboat from Penzance and Newlyn, the local committee decided to look for a better location for a boathouse which would allow the boat to be launched under any circumstances. They agreed on the need for a site that could provide a deep water slipway that could be used at all states of the tide and weather.

In August 1910, plans were drawn up by W.T. Douglass, Consulting Engineers of Westminster to accompany their report of July 29th showing a building 55ft long, with a slipway of 135ft from its frontage, dropping on a gradient of 1 in 4 to the low water mark.

That month, a letter from Thomas Holmes, Chief Inspector of

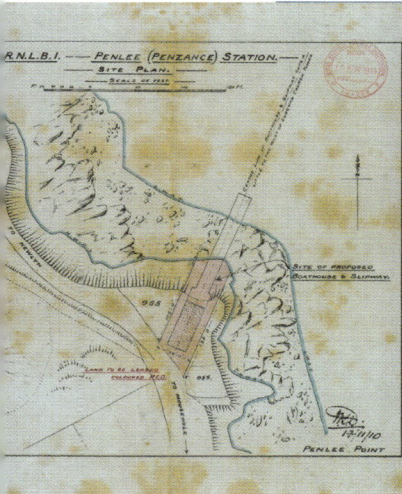

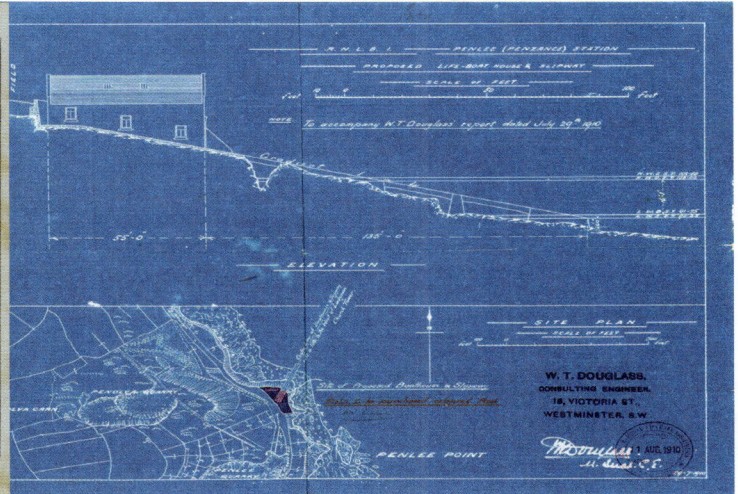

Lifeboats at head office sent to Colonel T. H. Cornish, Hon. Sec. Penzance Committee "approved the Engineer and Architect's recommendations to construct a masonry boathouse and reinforced concrete slipway at the site selected near Penlee Point". However, this required access through a field that was occupied by a market gardener, Richard Wright who, at first, refused permission to give up his tenancy. He finally held out for a 28-year lease at £2 per annum and £8 compensation, which was paid in March 1911, opening the way for the engineer and architect to proceed. There was also a proposal to build a 'dwelling house' for the use of the Coxswain, but this was never followed up.

In June 1911 advertisements for tenders were placed in the local and specialist press. Of the three tenders received, the committee recommended the (cheapest) one from J. Arundel of Bradford for the sum of £2850. A site meeting with Arundel's representative on July 21 set out the line of the slipway "with a view to starting work without delay". The first part payment of £300 was sent on November 8. Various tenders were also submitted for constructional ironwork and heating apparatus.

On December 16, Mr Douglass' Foreman of Works, Geo Greenway wrote to the committee: "Having been

On the very day the foreman went to observe heavy seas at Penlee, he witnessed the Norwegian barque Saluto II being driven ashore at Perran Cove. The Elizabeth and Blanche II rescued all 13 crew

Telegraphic Address. OCCULTED, LONDON.
P.O. Telephone Nº 99 Victoria.

W.T. DOUGLASS, M. Inst. C.E.
CONSULTING ENGINEER.

15, Victoria Street
Westminster
London 8th Nov 1911
S.W.

Penlee

Dear Sir,

The Lifeboat Institution will forward you tomorrow the sum of £330.0.0 for payment to Messrs. J. Arundel & Co. the Contractors for the new lifeboat house, etc. at this station, as a first instalment under their contract.

Yours faithfully,
Wm Douglass

Colonel F.H. Cornish
Hon. Sec

requested by you on a previous date to observe the state of the sea on the site of slipway in bad weather I now beg to inform you that during the heavy gale on Wednesday last [Dec 13] I took particular notice of the sea breaking on the site of the slipway and I am confident it was possible for the boat to have gone afloat at any time during the day. There were a great number of fishermen from Mousehole also present on the site during the time the lifeboat was going to the barque *Saluto* and they were confident the lifeboat would have gone afloat alright." Chief Inspector Holmes wrote to Col. Cornish "This only confirms the opinion expressed by your Local Committee when they recommended the removal of the Lifeboat to Penlee Point."

Approximately two years after work started on the building it was opened in October 1913 and was home to the *Elizabeth and Blanche II* until she was replaced in 1923. When the boat was launched for the first time from the top of the slipway it hit the water at such an angle and speed that the foredeck, and the dignitaries aboard, were completely submerged, and thereafter boats would be launched from part way down the slip. Although the boat could always launch from the slip, it was a different matter re-housing it in bad weather. There was no berth to moor alongside at Penlee and survivors, medevac (medical evacuation) patients and the boat crew would generally be landed at Newlyn, where the boat would be kept until weather conditions allowed it to be returned to the station.

Among the builders of the boathouse was George Webb Pomeroy, (seventh from left) who had the satisfaction of being on board the Elizabeth and Blanche II, *for its inaugural launch on October 22, 1914 (right)*

The Cox'ns during that period were: J. S. Brownfield (1913-1916) who attended two shouts, the first from the boathouse was one year after opening in October 1914, to the *Liguria,* and to the *Traly* in 1915; G. Dennis (1916-1920) who covered three shouts in 1917 and 1918 (SS *West Wales* and tug *Epic* twice) and Frank Blewett, Penlee Cox'n from 1920 right through till 1947.

When time came to replace the last of the pulling and sailing boats, the boathouse was modified to accommodate Cornwall's first motor lifeboat *The Brothers* (ON671). This was a new Watson class of boat, 45ft long, with a beam of 12ft 6in, and powered by the 6-cylinder Weyburn engine, built by J. S. White at Cowes, yard number 1566. The donors of the boat, the three Misses Eddy of Torquay, whose nephews they were commemorating, contributed to the modifications to the boathouse and slipway. A stone plaque outside the boathouse acknowledges this donation

During 1922, while that

work was taking place, the lifeboat was moored temporarily in the harbour at Mousehole. The first shout for the new boat came in early January 1923 to the SS *Dubravka*. According to The *Life-Boat* Journal: 'On January 3rd/4th, 1923 *The Brothers* lifeboat launched from Penlee Point following information received from the Coastguard that a vessel was in need of assistance near the Runnelstone Rocks. They found the SS *Dubravka* of Dubrovnik, with a crew of 31, at anchor, in a strong Westerly gale and heavy seas, having lost her propeller and dragging her anchors to within 20 yards of the Runnelstone Rocks. There was great danger as it was getting dark and the gale was worsening. The lifeboat took off 27 of the crew, and was so near the rocks that waves broke over both rocks and lifeboat.'

The original boathouse lasted from 1913 until it was adapted in 1922 to fit The Brothers, as seen here, and then there was a major reconstruction in 1930-31, to accommodate the W&S, with plans as shown below

The same year, also under command of Frank Blewett, the boat rescued 35 people from the *City of Westminster* which sank after taking the top off the Runnelstone. In 1924 it landed five survivors from the *River Ely* of Cardiff. In its time on station the boat made only six services, all of them to steam-powered vessels, landing a total of 67 people.

Pending the arrival of the next boat, 'our' *W&S*, plans were drawn up to enlarge and re-equip the Penlee boathouse. Architects Lewis and Lewis provided detailed drawings in May 1930. These showed a 7ft 6in extension to the lower level of the frontage, with existing doors lengthened and refixed. Two existing windows in the front gable were to be replaced by a single central window at a higher level to light the mechanic's workshop on the upper storey. A new 22 pane window was added above the doors in the new extension. This extension gave space for the longer 45ft 6in boat sitting further down the internal slip, which had a gradient of 1 in 4. The plans specify there was to be 40ft from the rear of the boathouse to the centre of the tipping cradle.

The width of the boat was the same as its predecessor, 12ft 6in, and the well measured 14ft at its top and 9ft at lower level, so no change was needed there. Additionally, a well was to be formed in the existing keel pit for the tipping cradle controller and a portable timber support for the keel. The existing drain under the winch platform was to be relaid. Removable timber bilgeways were added and concrete blocks set

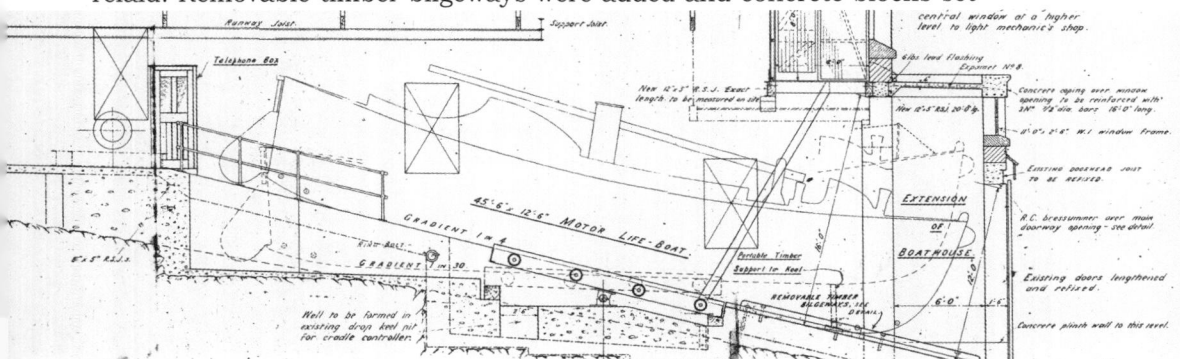

A still from the 1940 film S.O.S. following the crew members' arrival at the station's roadside gate shows the steep drop in level down to the entrance to the building. Note the date plate set into the gable BRITISH COUNCIL

into the walls of the well for building in ring bolts to hold the boat's restraining chains.

Presumably, the winch and its motor would have been upgraded at the same time as the utilities, such as heating, lighting and power supply to the workshop. A telephone box was erected at the top of the slip, and a standard roof vent installed to extract the boat's exhaust.

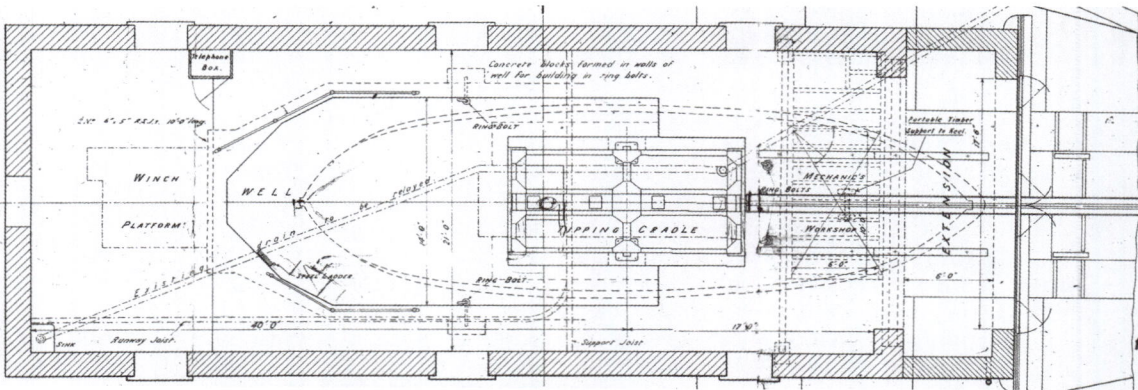

Mechanic John Drew (left) and Cox'n Eddie Madron beneath the boat, with tipping cradle visible. And (right) one of the service boards hanging in the boathouse, covering the early years of the W&S. The service at the bottom should actually be credited to the relief boat B.A.S.P.

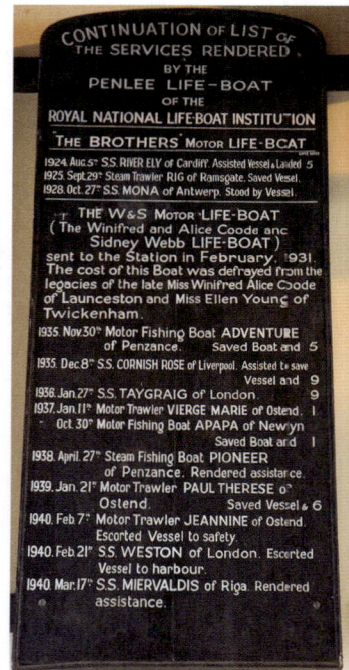

CONTINUATION OF LIST OF THE SERVICES RENDERED
BY THE
PENLEE LIFE-BOAT
OF THE
ROYAL NATIONAL LIFE-BOAT INSTITUTION

THE BROTHERS MOTOR LIFE-BOAT
1924. Aug 5th S.S. RIVER ELY of Cardiff. Assisted Vessel & Landed 5
1925. Sept 29th Steam Trawler RIG of Ramsgate. Saved Vessel
1928. Oct 27th S.S. MONA of Antwerp. Stood by Vessel

THE W&S MOTOR LIFE-BOAT
(The Winifred and Alice Coode and Sidney Webb LIFE-BOAT)
sent to the Station in February, 1931. The cost of this Boat was defrayed from the legacies of the late Miss Winifred Alice Coode of Launceston and Miss Ellen Young of Twickenham.
1935. Nov 30th Motor Fishing Boat ADVENTURE of Penzance.   Saved Boat and 5
1935. Dec 8th S.S. CORNISH ROSE of Liverpool. Assisted to save Vessel and 9
1936. Jan 27th S.S. TAYGRAIG of London.   9
1937. Jan 11th Motor Trawler VIERGE MARIE of Ostend. 1
    Oct 30th Motor Fishing Boat APAPA of Newlyn. Saved Boat and 1
1938. April 27th Steam Fishing Boat PIONEER of Penzance. Rendered assistance.
1939. Jan 21st Motor Trawler PAUL THERESE of Ostend.   Saved Vessel & 6
1940. Feb 7th Motor Trawler JEANNINE of Ostend. Escorted Vessel to safety.
1940. Feb 21st S.S. WESTON of London. Escorted Vessel to harbour.
1940. Mar 17th S.S. MIERVALDIS of Riga. Rendered assistance.

58

Launching the Boat

'Knock her out!' That crucial moment which all crewmen look forward to: The Launch. The famous hammer was used to knock out the pin on the shackle holding the chain on the tipping cradle which let the boat thunder down the slipway. For 38 years the procedure was controlled by the winchman Raymond Pomeroy Sr. (right) later followed by his son.

Rehousing the Boat

Contact is made with the shore astern of the boat and the undersea warp is fished up forward. This enables the boat to be aligned with the centre channel of the slip and the crew are hauling the boat into contact

Once firmly on the slip preventers are rigged to hold the boat in place while the winch wire is hooked to her stern and the other tackle dropped over the side of the boat

Rehousing the boat calls for experienced hands and it can be dangerous. In 1962 when rehousing the *Solomon Browne* 78-year-old Jimmy Pentreath was killed when the wire caught his legs, and 72-year-old Bobby Blewett was knocked unconscious in the water. He survived, but since then a 65-year age limit has been imposed on shore staff and boat crews

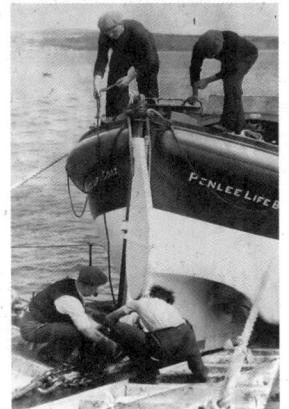

The boathouse was opened for business in 1913, housing the *Elizabeth and Blanche II*, followed by *The Brothers* in 1923. It was adapted and enlarged to take the *W&S* which arrived in 1931 PHOTO: HARRY WELBY

FLANK STATIONS: The Lizard

The third boathouse at Church Cove on The Lizard was built in 1914 and was in use until 1960 when it was replaced by a new station on the eastern side of the peninsular. The inset photo shows crowds watching lifeboat movements in the 1950s

Throughout the RNLI, neighbouring lifeboat stations are known as 'flank stations'. The two stations adjoining the Penlee boat's territory are The Lizard to the east and Sennen Cove to the west, respectively the most southerly and most westerly lifeboat stations on the British mainland. In earlier times, up until 1929, there was also a station at Porthleven, then the UK's southern-most commercial port, and the Lizard station was originally sited in Mount's Bay on the west of the headland at Penleor, before moving to the eastern side in 1961. For a short while from 1867 to 1908, the RNLI also operated a station at Mullion. In recent times there has been a small inshore boat operating from Marazion during the summer holiday season only.

The Lizard Peninsula pokes its gnarled serpentine finger into the often wildly swollen waters of the Atlantic Ocean as they are funnelled into the English Channel, exposed to all the extremes of the elements and vicissitudes of nature. It can be a very bleak environment on land, and a terrifying hazard when approached by water. The turbulent waters around this headland are amplified by overfalls as the tides race. The passage from east to west into Mount's Bay can be lumpy, especially when winds are Nor'westerly against a flood tide which creates short, steep seas. On Spring tides there can be a 3 knot increase and a tidal range of up to 5 metres. Prudent navigators would stay at least a mile or two offshore.

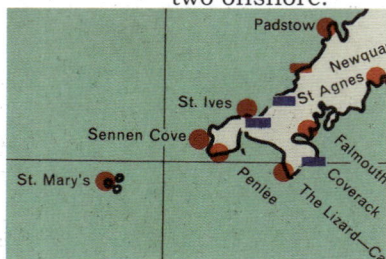

The Lizard is the home of the southernmost human settlement in the British Isles, and as quoted by one of the successive Sir John Killigrews: 'most of the houses were built with the ruins of ships'. In the 19th century the Coastguard established a base here, later augmented by the Cliff Rescue/Life Saving Apparatus (LSA).

The first lifeboat station at the Lizard was opened in 1859 above Polpeor Cove on the west side of the headland. This proved to be a difficult site from which to launch and retrieve the boat, and it was not possible to operate a larger lifeboat there because of the shallow water at low tide, and the impossibility of hauling a heavy boat up the beach. Recovery at Polpeor could be a dangerous operation in a South Westerly gale, and the approach, via a narrow channel between rocks, was tricky to navigate at night. Thus, in 1885 another boathouse was built further up the High Water mark in the sheltered Church Cove where the sea was often calmer and there was deep water at all states of tide, making the launch and recovery of a larger lifeboat easier. A third boathouse was constructed there in 1914.

The first motor lifeboat on station was the 38ft *Frederick H Pilley*, delivered in 1922, followed in 1934 by the 40ft Watson class *Duke of York*. Eventually, the station was moved around the corner to the east side of the headland and the lighthouse, facing away from the prevailing South-Westerlies. Construction of the new (now the old) Lifeboat Station at Kilcobben Cove started in 1959 and a new 52ft Barnet class boat arrived in October 1960. This boat was 12 foot longer and twice the weight of her predecessor. The station was officially opened in July, 1961 and the lifeboat named *The Duke of Cornwall* by H.R.H. The Duke of Edinburgh. From that side of the headland, lifeboats were better placed to respond to outbound vessels approaching the Lizard from the east, which had the notorious Manacles reef to avoid. On July 23, 1962 a launch by the Lizard lifeboat was featured in the first television programme to be transmitted live to international audiences via the Telstar satellite from nearby Goonhilly. In 1963, the Lizard station was combined with the Cadgwith station a couple of miles to the north east, and a completely new boathouse was built in 2010-2012 to accommodate the latest slipway-launched Tamar class boat.

The Life Saving Apparatus (LSA) building at The Lizard, known as Rocket Cart House, was constructed in 1869. In 1872 Lloyd's of London, underwriters for maritime insurance, set up a signal station at Bass Point (Beast Point) which housed an office for the GPO telegraph to announce the sighting of inbound ships heading through the Western Approaches to the Channel and northern European destinations. Many of those vessels would be aiming for Falmouth as first port of call; that fabled destination was literally just around the corner from the Lizard, but first those turbulent waters had to be navigated safely.

Sennen Cove

The station to the west of Mount's Bay is Sennen Cove, established in 1853. The boathouse is situated just one mile North East of Land's End at the western extremity of the British mainland. The tiny village, and its wide, flat beach, is protected to some extent from the full force of the Atlantic Ocean and the frequent westerly gales by the high granite cliffs of Pedn-men-du at the southern end of the bay, but the exposed location offers some of the most difficult conditions for lifeboat operations anywhere in Britain. There have been frequent occasions when the Sennen boat has been unable to launch into the breakers and the Penlee boat has to be called out to cover. The two stations have also worked together on several

TO THE RESCUE   SENNEN LIFEBOAT   HO

of the more arduous shouts. Sennen Cove is the only RNLI station to have two operational slipways; one that is sheltered by the breakwater is used for recovery at high tide, while the launching slipway is used at low tide. As well as Penlee, the other flank stations are St Mary's on the Isles of Scilly to the West and St Ives to the North.

The RNLI built the first boathouse at the top of the beach following the wreck of the brig *New Commercial* on the Brisons in 1851. On passage from Liverpool to Santa Maria (the Spanish Main), the vessel struck a rock between the Great and Little Brisons, about 4 miles North of Land's End near Cape Cornwall, in a SSW gale. The nine crew, plus Captain Sanderson's' wife, were stranded on a rock ledge overnight, only to be washed off in the morning. Just three of them survived: a seaman who came ashore on a plank of wood, and the Captain and his wife, who managed to reach the Little Brison. During the day a Revenue cutter tried to rescue them with a rocket line, while a crowd reported by *Lloyd's List* to comprise *five to six thousand people* spectated from the shore. There must have been several charabanc excursions arranged to transport that many witnesses. The couple were saved but, sadly, the wife expired before she reached land. Medals were awarded to the cutter crew and their commander Captain Davies, and rewards given to several local fishermen, and the following year the National Shipwreck Institution delivered a 25ft Peake class lifeboat to the community.

The boathouse, built in 1853 to accommodate that boat, which made only one service, was extended in 1864 to fit a larger 33ft self-righting lifeboat *Cousins William and Mary Ann of Bideford*. Two more self-righting boats followed, the 34ft *Denzil and Maria Onslow* (1880-1893), and the 35ft *Ann Newbon* from 1893 till 1922. In 1874 a new boathouse was built across the road, but in 1896 that was replaced by another

The Susan Ashley *being rehoused bow first in 1968.*
PHOTO: DEREK HARVEY

building on the original site. A new slipway was constructed in 1919 and the first motor lifeboat, *The Newbons,* arrived in 1922, when the boathouse had been fitted with a turntable to rotate the boat after its being hauled up the slipway bow-first. Ten years later a second, additional, slipway was provided. *The Newbons* stayed on station until 1948, when it was replaced by the 41ft Watson Cabin class *Susan Ashley,* until 1973.

There are no straightforward currents around the Land's End peninsular, according to the ex-Penlee Mechanic Johnny Drew, talking in the mid 20th Century: "How should I describe it? There's a six-hour tide: six hour flood, six hour ebb. There's no what we call 'winding' tide, it's direct. High water flood tide, tides going North East. When it's low water ebb tide, it's South West. Right opposite, North East and South West. The wind makes no difference, it's the tide. In your former days when you were on your compass, you would know your currents and how they would run. It's incredible really because the Pilot books that I have give you all the marker buoys around the coast, and the waterways. Providing if you come in and carry out what's in front of you in them books you can't go wrong."

As we will see in the following pages, not every ship's navigator was so well informed.

The Boys on the Boat

FOR generations lifeboat crews would be called into action by the maroon, a loud rocket that reverberated around the waterfront and showed a bright flare out at sea. These were withdrawn for 'safety' reasons in the early 2000s, once the boat crews had got used to being contacted on individual pagers. "When the maroons for the lifeboat went up everybody in the village, everybody who could heave a leg, went — whether you were young or old," said Johnny Drew, the mechanic who spent 29 years on our boat.

❝It wasn't like the sailing days, because the boats now were mechanised and they had reduced crews of seven. But you could have three or four crews for the number of people that ran to volunteer. But only seven could have the lifejacket. You had to take a lifejacket and get on board to become one of the crew. Then it was really who the Cox'n wanted to take. If he thought I was better able than you he would say 'I'll take Johnny'. But there was a lot of controversy at the time because there was a lot of jealousy, but that was part of the job. We, as youngsters of 15, 16, 17 didn't think we were man enough for the boat. But we would run to volunteer. I've seen elderly men, men that we would term old now, had to run to Penlee — no cars then. Had to run. I've seen elderly men out of breath falling by the wayside. They couldn't do no more, but the youngsters would go and leave them. Many occasions I've seen that. Old men did their best but couldn't do any more. [The run from Mousehole to the Penlee Point is about one mile and it is uphill.]

Go over the steps and the boat house would be full [of runners]. There would be enough men there to form three crews. But gradually it diminished over time because the local inhabitants were dwindling and there were not too many local people, not so many seafarers. Fishing families were dying out, visitors coming and buying up cottages and coming down to live.

The maroon was the recognised signal for calling out the lifeboat. In them days there were far more shipwrecks on the coast because they were sailing ships and they would be caught on the deadly shore in the bays. When

Young and old would run to bag a place on the crew when the maroon was fired. Cox'n Blewett leads the way

BRITISH COUNCIL

the weather made what we call 'beat out', they couldn't get out, therefore were trapped in the bays. As seen in some of the old photographs, you would see these big sailing boats all dismasted and wrecked along the Cornish coast. They were all caught in this weather, whereas if they were mechanised they might have steamed out and got clear. But not in that day.

A lot of people don't know, but the firing of the maroons was for three purposes. One, to call out the lifeboat crew; two to notify the public of the lifeboat; and three, if the casualty was in range of hearing the signal, they knew the lifeboat was on the way. That was three reasons for firing the maroons. They were terrifically loud. The communities in them days would jump like a flash, jump at everything in your younger days, see who could get there first. The first time you went, you was dying to go [on a shout]. And when you did go, you felt like a giant. There was a certain amount of pride to be a lifeboat man. Many times I've been out, I don't know about the rest of the crew, but many times I've offered up a small prayer for my safety. And you had to have courage as well as pride.

In them days crews were made up of men who were skippers of their own fishing boats. So you could have every member of that lifeboat crew capable of taking charge of the boat. Their knowledge was fabulous really. They knew everything that happened, handling the boat and the courses and the compass, which was the mainstay. The compass was everything in them days. They had the courses, they had everything. They could consult each other, see, seven men on a lifeboat crew. All those men were capable, and some perhaps a little better than others, but they were all capable of taking charge.

When the maroons went you would run to the Lifeboat station out of excitement. But you were always waiting for the day: What wouldn't I do to get on the crew. At last the day arrived when you got there. It was an urgent call.

Being young you got there quicker than someone older than you. The main thing you did when you got to the lifeboat house was to run to the rack and take a lifejacket. That gave you a place on the boat. This was normal procedure. Take not your oilskin, take your lifejacket. So you take the lifejacket and jump aboard hoping that this was it. Well, that's what happened to me on my first occasion and I felt like a giant. Just hoping and waiting because the Cox'n was entirely right in what he wanted. He had his own idea, he wanted the best crew each time he went. When I was a youngster and wanted to go on the first occasion I might think I was alright, but by some coincidence I got into the boat perhaps a little quicker than someone who was much better than me as a crew man. Sometimes the Cox'n would say to me, 'Look my son, wait till next time. I'll take so and so this time, perhaps you next time.' It was to the Cox'n's discretion who to take. As you get older and with experience you understand, whereas when you are younger you just jump at everything, but when you got older you realise he was right. He wanted the best crew. That's what it was like until I got my first call and then I felt big as a boat. It wasn't so bad weather, but it was to a ship. It was the *Dubravka*, I think. She had gone aground in St Loy Bay. There were no lives lost.

You had numerous calls in them days when I was young in the 1920s, because if it was foggy weather ships didn't have radar or nothing then. They just went by a course on the compass and if they were slightly off course they would be ashore, and if something unforeseen happened then they were wrecked. The navigation equipment wasn't nothing compared with what it is now. It's all electronics now, whereas then the captain of the ship had his chart and compass and that's all he went by. If his compass might have been out, or for some other reason, then disaster came and therefore the lifeboat was called.

Edward 'Ted' Downing (grandfather of Cox'n Patch Harvey) running through Mousehole to the boathouse in the film S.O.S.
BRITISH COUNCIL

Conditions on board

Each lifeboat carried its own rations, which in the early days was sealed tins of chocolate, sealed tins of biscuits and a bottle of rum. Some Coxswains were more strict that others. My first Cox'n [Frank Blewett] was very strict on the opening of any of those rations. Sometimes the men in the sailing days suffered great exposure, different from later days. They didn't have any protection, and sometimes a tot of rum would have warmed their system, while still being wet through. But that was left to the jurisdiction of the Coxswain. If a man was dying of cold he couldn't have anything unless the Cox'n gave permission. So the sealed rations would not be opened. Well, that all depends on what type of man the Cox'n is, doesn't it?

The early boats were all open. The *W&S*, what they call the 45ft 6in Watson Cabin type, had a cabin below decks for perhaps 10 survivors. But just a small canopy on top for protection of the Coxswain and mechanics. The only protective clothing was what you stood up in, other than an oilskin and a lifejacket. You had no boots supplied until the later years, I suppose 15 or 20 years later. For instance if the lifeboat was called, other than if you had a pair of rubber boots at home and you knew it was a rough night, you normally ran in your ordinary shoes or boots. And when you got to the lifeboat station you took your lifejacket and oilskin and your sou'wester: three things. You put them on and that was all the equipment you had. In fact, you wouldn't be launched more than minutes before you were wet through, because once your legs were wet you were soaked. That went on for abut 20 years, but in later years each man kept his own pair of boots at the station until the Institution issued them. I know the Penlee had some Ladies Guild, in the 1930s, when the *W&S* was there, they knitted a lot of head scarves with the initials 'Penlee lifeboat' on them. Long white scarves that you could wrap round your neck. In later years the Ladies Guild were contributing a lot of materials to the crew, quite a lot.

In addition to the chocolate, biscuits and rum, there were tins of hot soup. They had the self-heating cans first, but now the non self-heating cans because they've got a stove on board so you can heat them. You were allowed to smoke on board and some of them would be chain smoking. Always had tins of cigarettes, in later years, as your ration from the RNLI. Along with the rum and chocolate there was the tin of cigarettes. One gentleman [Russel Snook] use to come here and he got to know us well. He said he would keep the lifeboat supplied with cigarettes. I would get them from the tobacconist and he opened an account there. Had two tins at a time. So that was always a reserve.

During the war, on one or two occasions up in the East coast somewhere, the lifeboat had been called out to rescue German airmen who had been shot down. The Institution realised that they were always armed. Apparently one lifeboat picked up two German officers, one of whom threatened them with a revolver. So then the authorities realised the danger lifeboatmen were in; they could have been shot. The Germans could have taken the lifeboat and gone. From then on, all lifeboats carried two Lee Enfield rifles. Every station was supplied with them, to be stored in the cabin. It was the mechanic's job to keep them clean and in readiness. This was at a time when you knew nothing about armoury but it was something else you had to adapt to. There was no protective clothing provided, only ARP [Air Raid Patrol] tin hats. We didn't have service gas masks, they were never issued, but we had our civilian gas masks.

Working hours

All the time I was in the lifeboat service I was on call 24 hours a day. I would never normally go to Penzance but every Saturday I had to report to the Hon. Sec. He insisted I did. Then I had to get my wages but he insisted I should visit him just for him to hear me say 'Everything at the station is OK'.

If I left home the wife would be home. If I went to Penzance and I left the Hon. Sec.'s to go to two shops I would let her know which shop I would be in. I remember on one or two occasions I was called out from Penzance. They weren't urgent calls at the time, but I wouldn't take any chances. Once or twice I stopped at a cafe there to have my lunch. Then another occasion I was up paying my electric bill, I told him where I would be. One day after the war, I had been at the lifeboat house all morning doing normal duties. Went in to lunch and then I had to go to Newlyn for some reason, but during that time I was going to stop at the lifeboat station to switch the batteries on charge, or something like that, and then catch the 2 o'clock bus from there as it passed. So I goes to Newlyn, walks down over the pier and one or two chaps are there talking on the trawlers. I

went aboard a boat and was just going to do something, took two spanners to attend to this job. I just had the spanners in my hand and somebody shouted 'Johnny, your wife's rang up. Lifeboat!' I dropped the tools and run up the pier and I said 'Any cars about?' And one of the chaps who I knew had a big motorbike and he said 'Jump on boy, I'll give you a lift. I jumped on that bike and going over Newlyn slip I was airborne. I gets to the lifeboat station thinking the crew would be there but when I get there I'm the first. Get my keys and unlock. I had to ring up the Cox'n and tell him I was out there and when I rang my wife to let her know she said 'Hurry up and get out there.' I said 'I'm here'. And that was for an aircraft that ditched off Cudden Point and we launched. But the helicopter went out and picked them up just before we got there. That's what it was like. I could pinpoint to my wife exactly where I would be. I never went to Penzance or Newlyn unless they knew where I was. **"**

The Cox'n and Johnny both had telephones provided by the RNLI. Nim Bawden, a shipwright and crew member who would later become Johnny's assistant mechanic, recalls that Johnny could afford to live on the Mechanic's wages of £14 per week because he also owned the *Bonnie Lass* pilchard driver, and he had a house provided by the RNLI when he was promoted to Motor Mechanic. Nim, whose regular employment was shipwright, was a volunteer second mechanic earning £1 per week for nine hours. At the time the boat crew earned 30 shillings per launch, then 10 shillings per hour after three hours. Sometimes there were rewards, maybe from the shipping company if they'd done a medevac. One time they received £18 per man, which was a lot. That day Nim was called in to the office by Cox'n Jack Worth with the good news: "We've got a puff half." [An unexpected bonus, or 'perk']

As well as the boat crew, the lifeboat service depended on the shore team which was led by the Honorary Secretary (Hon. Sec.), later known as Lifeboat Operation Manager (LOM) and the team of launchers, winchman and signalman. During bad weather when the boat had to remain overnight at Newlyn there would also be a watchman left on board. Backing them all up were the local Branch Committee, the Ladies Guild and a legion of fundraisers from all over the country.

At the Helm of the Boat

The word 'Coxswain', meaning officer in charge of a ship's boat, or cockboat, is derived from the old French *coque* 'cockle shell' and the old Norse *swain* for 'boy, servant or hand'. In the RNLI the Coxswain (or Cox'n) is classified as an Officer, in command of the boat, reporting to the Hon. Sec. and the Branch Committee.

During the 29 years that the *W&S* served at Penlee there were just three Coxswains in command: Frank Blewett from 1931-1947, Edwin F. Madron from 1947-1957, and John T. (Jack) Worth from 1957-1960.

The *Regulations of the RNLI*, 1950 edition, state the Coxswain is responsible for the 'efficiency, good order and cleanliness of the boathouse and slipway, the lifeboat and her gear. He will have one of the keys to the boathouse. He must see that the lifeboat is at all times ready in every respect for a launch on service [. . .] He will be in charge of all operations connected with the launching, hauling up and housing of the lifeboat. He will be in charge of the lifeboat when she is afloat, and may not, unless incapacitated, give up the command to any other person [. . .] He is not responsible for the technical operation of the lifeboat's machinery, which is under the charge of the Motor Mechanic.'

Along with the Motor Mechanic, the Coxswain of an all-weather lifeboat (ALB) is often, but not always, a full-time professional. (These days, many stations employ a single coxswain/mechanic.) The bulk of the crew are volunteers, among whom there would also be at least one qualified Second Coxswain, although many of those volunteers would be fully capable of taking charge of the boat as acting Coxswain, if needed. In the early days of the RNLI almost all crew members would have been fishermen or other types of mariner, with a high degree of competence in boat handling. RNLI Coxswains have always worked their way up to the 'top hand' position after many years as volunteer crew men. The Penlee coxswains and mechanics during the time of the *W&S* were all from established families of fishermen.

Frank Blewett, Cox'n 1931-1947

The first of the Coxswains to take charge of the *W&S* on the boat's delivery in 1931 was Frank 'Nailer' Blewett, the longest serving skipper of our boat, who had been the station Cox'n at Penlee since taking

command of the *Elizabeth and Blanche II* in 1920, followed soon after by *The Brothers*. Before that he had spent four and a half years as second Cox'n and two and a half years as bowman, since joining the crew in 1913 on the opening of the Penlee boathouse. In his lifeboat career he was credited with saving 149 lives aboard the three boats (plus relief vessels) he commanded, and was recipient of the Institute's Bronze medal for bravery in 1936, following the *Taycraig* service on the *W&S*.

Born in 1885, Frank was a fisherman from a Mousehole family, who owned the pilchard driver *We'll Try*, for which they never had trouble finding a crew. He first went afloat on the Newlyn pulling and sailing lifeboat in 1912. Later described by a crew member as a strict superior, and with fixed habits, he was teetotal (apart from taking a Christmas tipple) and was sparing with his praise or compliments. His grandson, also Frank, described 'Nailer' as a very gruff man who didn't waste time on pleasantries. He was passionate about his lifeboat but he didn't talk very much about his work or the services — he was a humble man, but also quite a hard man. He was totally dedicated to the lifeboat Institution and he did enjoy giving trips on the boat to visitors.

Frank Blewett's wartime permit to travel and enter protected places under defence regulations

Before the arrival of the *W&S*, Cox'n Blewett was involved in some of the most dramatic rescues of the early 20th Century, including the service on *Elizabeth and Blanche II* to the steamship *Concordia*, which had lost its propeller near the Wolf lighthouse in 1922, for which Blewett received a Letter of Appreciation from the committee. At the end of that year the station's first motor powered lifeboat *The Brothers* arrived, and two weeks later, in January, 1923, she attended the steamer *Dubravka*, with a lost propeller and dragging her anchor close to the Runnelstone rock, taking off 27 survivors with waves breaking over the rocks and the lifeboat. In October that year the SS *City of Westminster* did, infamously, strike the rock, taking the top off it and rendering the ship a total wreck. That night Frank and the crew of the *The Brothers* saved another 35 lives.

When the *W&S* arrived at Penlee, noting the boat had the luxury of a cabin for survivors, Coxswain Blewett declared: "When we have snatched a man out of the jaws of death this is where he begins his second life." During his 16 years as Cox'n of the *W&S*, Frank Blewett saved many more lives and witnessed several tragic events, described in subsequent chapters, proudly upholding the lifeboatman's tradition of service to fellow seafarers. But Frank was more than just a Coxswain; in the 1930s he was made an Inspector Coxswain. After having written a very long letter to the RNLI pointing out how useless their Inspector was the RNLI promoted him to Insp/Cox'n, meaning that he didn't have to be inspected again! He sometimes signed reports 'Superintendent Coxswain'.

His greatest moment of glory came in 1936 when he was awarded the Bronze Medal for the rescue of nine crew members from the steamer *Taycraig*, driven onto rocks in Mount's Bay in a gale in January of that year. In June 1936, he was one of the record number of 15 lifeboatmen from England, Ireland, Scotland and Wales invited to London to receive their awards. The men were treated royally in town, first being taken to the Palladium music hall, and on the next morning posing for press photographs in their oilskins and sou'westers in the gardens of Life-boat House in Victoria. In the afternoon they were decorated by HRH the Duke of Kent, and then taken to the House of Commons to be entertained to tea. In the evening they were the guests of the Coliseum theatre, to watch the musical comedy 'Twenty to One', where they were seated in the Royal box and five other boxes. At the end of the performance the impressario Mr Lupino Lane introduced the lifeboat heroes to the audience. The spotlights were then turned on the boxes and the lifeboat men acknowledged the audience's applause. Frank Blewett retired in 1947, aged 62, as did the 2nd Cox'n Luther Oliver. That was just a few weeks before the *Warspite* incident, and Frank was reported to be seething with envy that he hadn't been on one of the Penlee station's most famous shouts.

One of Frank's great passions was to acquire, learn and use new, unusual and strange words. He was a Pupil Teacher at Mousehole School. He smoked Woodbine cigarettes, but never in the house, and he was not a Chapel goer. After retirement Frank went back to fishing with his boat the *Daffodil*. His grandson recalled him never driving a car, but first thing in the morning he would make a daily round in his boat to Newlyn fish market, followed by breakfast at home and then a full day fishing for mackerel off Porthcurno.

Back in 1910, Frank's sister Janie had married Adam Torrie, a survivor of the *Baltic* shipwreck, who had subsequently lodged with the Blewett family. At their wedding was her nephew, one Eddie Madron Jr. Frank died in 1973 aged 88, and is buried at Paul Church. He was a cousin of *Solomon Browne* crew members, Barry Torrie (Adam's grandson), Stephen Madron & John Blewett.

Edwin 'Eddie' Madron, Cox'n 1947-1957

Edwin F. Madron joined the lifeboat crew in 1925, and from 1934 to 1947 he served as Second Coxswain to Frank Blewett, whose role of Cox'n he would inherit when Blewett retired. During World War 1, Eddie had served in the RNR on the battleship HMS *Canopus*, in campaigns from the Falklands to the Dardenelles. In WW2 he was called up again as a Chief Petty Officer, while still doubling as a lifeboat volunteer. His first shout after promotion to full-time Coxswain in 1947 was the highly dramatic, and well-publicised, service to the battleship HMS *Warspite*, which earned him the Institute's Silver Medal, along with a Bronze Medal to Motor Mechanic John B. Drew, for their services rescuing a skeleton crew from the obsolete warship that was wrecked in Mount's Bay, while being towed to a shipbreaker's yard. Edwin also received the Maud Smith award for the bravest deed of that year and the rest of the crew received certificates. This dramatic episode is covered at length on pages 137-143.

Like most of the Penlee crew, Madron was a fisherman from the village of Mousehole. He ran his own trawler, the *Renovelle* (PZ107), which in the late 1930s voyaged up the east coast of Britain in search of fish. On one trip in 1938 the vessel landed 75 crans of herring at Lowestoft. A cran was a measure of

Sorting the 'Silver Darlings' on board PZ107 in Mousehole harbour. Eddie Madron is far right, along wth lifeboat crew men Ned 'Strags' Tregenza (at back with fag in mouth) and Jack 'Slim' Wallis, second from left

NIM BAWDEN COLLECTION

uncleaned herring, equivalent to one standard box of about 37.5 imperial gallons — roughly averaging 1200 fish.

When war was declared again in 1939, he and the *Renovelle* were requisitioned for coastal patrol duties. He was asked to run secret agents across the Channel to German-occupied France but the senior Naval officer at Penzance, Lt Commander Wells, blocked that, as "It would have been suicide to let *Renovelle* enter enemy waters". Commander Wells later recalled that one night he was on leave, having a quiet pint in his local when the maroon went up. "Madron dropped his pint and ran like a stag to the lifeboat. A trawler had been sunk in Mount's Bay and they picked up survivors."

After the war, job opportunities were scarce and fishing was poor, so Eddie's promotion to Cox'n was timely, guaranteeing a regular salary and a cottage. As well as his services aboard the *W&S*, Eddie was involved in some dramatic shouts using relief boats, while the regular boat was away for maintenance. There were several calls with the *Millie Walton* to the ongoing *Warspite* salvage and, in 1956, the steamer *Yewcroft* came to grief in a dramatic incident, breaking its back in the same area.

It it a cliche to say the RNLI is like one huge extended family, but in many stations it is common to find close blood relatives serving concurrently and/or consecutively. The Madrons are one such family: Edwin's father Joseph had drowned in Plymouth aged 40, when he was believed to have fallen off his lugger the *Velox* of Mousehole. The father of seven children, he had been chosen to be lifeboat Coxswain when the boat was moved to Penlee. By tragic coincidence, one of Eddie's own sons, Edwin Francis, aged 24, also drowned in Plymouth when he attempted to save his younger brother Joseph who had fallen overboard from their father's boat. Edwin Jr and elder brother Jimmy jumped in and the young Joseph was saved. Joseph served alongside Eddie as 2nd Cox'n and mechanic, often with his own sons Eddie and Abraham. Confusingly there were two generations of Madrons named Edwin F. and two also named Joseph. The crew list for a shout on the relief boat in July 1950 included E.F. Madron, J.J. Madron, J.B. Madron, E.F. Madron Jr and J.J. Madron Jr

Eddie Madron Senior retired in 1957 after 32 years on the Penlee crew: 10½ years as Cox'n, 12½ years as 2nd Cox'n, 9 years on crew. Awarded the Coxswain Certificate of Service and an annuity. Joseph retired in the same year after 1½ years as mechanic (and 1½ years at Mumbles) 2½ years as Mechanic, 9 years as second Cox'n, and 9 years in crew, after helping to save a total 134 lives.

MFV *Renovelle* also worked for a while as a Trinity House tender on the Wolf Rock lighthouse. Eddie Madron, his eldest son Jimmy Basset Madron and Borrey Harvey worked on the erection of a new upright crane used for heaving stores ashore.

The young Eddie Madron in his naval uniform during service in the First World War

75

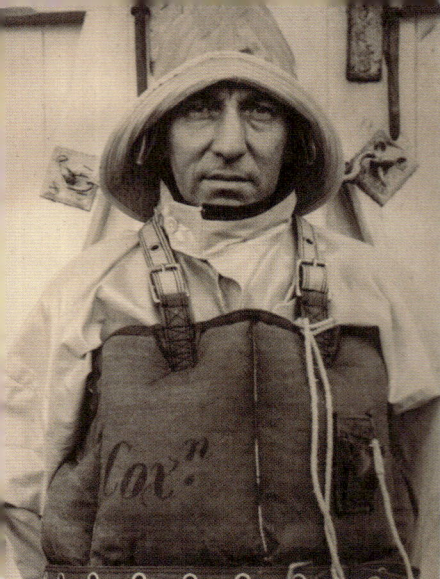

John T. 'Jack' Worth, Cox'n 1957-1970

Jack Worth, the third and final Cox'n of the *W&S* at Penlee, had a career that spanned three eras in lifeboat evolution, serving on *The Brothers*, the *W&S*, and finally the *Solomon Browne*. As a crew member, Jack experienced some of the most stirring episodes of lifeboat activity. Like his predecessors, Jack was a fisherman's son, who joined the lifeboat crew in 1923 at the age of 17. He was bowman for six years, second coxswain for two years and served as Penlee Coxswain from 1957 to 1970.

Jack was born in 1906: "I remember Mother saying I crawled (too young to walk, see) in from there, where we lived, and she missed me and when she found me I was down in the harbour! I was always in punts, always. Mad for the sea. I can remember back to about six, seven; the *Elizabeth and Blanche II*, the pulling and sailing lifeboat, brought from Newlyn to Penlee, to the new boathouse. That was 1913. And I was thinking and wishing: 'Oh! I'd like to be aboard of her!' and every time the maroons went, I always ran up to the Lifeboat Station!"

Jack first went to sea on fishing boats, including those of his father and uncle, followed by a few deep sea voyages on luxury yachts. During World War 2 he served as a Petty Officer Cox'n and later ran two fishing boats of his own, the *Leader* and the *Porth Ennis* (PZ39), with which he became closely identified. In July, 1958 he made a lifeboat service in the *Porth Ennis* to assist an open boat spotted in difficulties off Penzer Point.

His time in charge of the *W&S* was brief, inheriting the Cox'n role only three years before our boat was to be replaced by the *Solomon Browne*, but he had been on the crew since 1923 and was second Cox'n for a year, so he knew the boat intimately. He then served another 10 years on the *SB*, with Johnny Drew and Nim Bawden on the engines. He experienced some of the station's most dramatic services. The first serious

As a lad Jack went to sea on several luxury yachts including the Sunbeam, owned at the time by Lord Runciman, and voyaged to the Med as far as the Greek islands. In the crew photo above right he is seen second left in the back row

drama to occur on his watch as Cox'n was the grounding of the *Pluie de Rose* in April 1959.

Nim Bawden, who served with him as Mechanic, says Jack was a good Cox'n and a great navigator. One time when they were out looking for the Navy sail training ship HMS *Temeraire*, which was missing in foul weather, Jack worked out his position from the tides and how long he'd been out. When asked by a Navy minesweeper if he knew his position he replied. "No, I don't know where we are but I do know my way home." He was the one who taught navigation to Trevelyan Richards, the next in line to be Cox'n of the *Solomon Browne*.

Jack was the most outgoing of our boat's Cox'ns, and he and his crew would happily take members of the public out on trips on the annual Lifeboat Day. Friends and family members would get to ride the boat down the slipway as if on a fairground water-splash ride. Then the general public would be invited to board from the harbour wall. Sometimes 40 or 50 people would cram the deck.

Jack did like a drink, but never on board the boat, and he wouldn't let the crew open the rum when on service. Only once, Nim remembers, on the *Solomon Browne*, after standing by the tanker *Firth Fisher* during heavy weather in 1969, shortly before Jack retired, he did break out the rum. "Stream the drogue and give the boys a drink. Save one for me." At the local pub Jack would wear his RNLI jumper back to front. As he told the bar staff: "Well *you* know I'm on the lifeboat but the visitors [behind me] don't know that." Jack was always jovial when he manned the lifeboat stand at the London Boat Show and he regularly attracted the young Aer Lingus stewardesses, who would keep him supplied with Guinness.

Jack was a well-known resident of Mousehole where he took up painting after his lifeboat career

Crew mates recall that although he was strict and he could bark out loud, you could enjoy a bit of banter with Jack. He was involved with various local activities such as the football club, harbour sports, the fishermen's co-op and the campaign to establish a lighthouse at Tater Du. And he was an avuncular presence on the Mousehole waterfront. Once, when the young Elaine Bawden had saved up to buy herself a rubber dinghy to paddle about in, Jack asked to have a look at it and immediately pulled out his pocket knife and punctured the inflatable — and he would do the same to any young holiday maker foolish enough to risk floating out to sea. Elaine already had her own punt to boat about in the harbour with full confidence, which Jack approved of, but he believed these newfangled rubber dinghies were dangerous playthings.

On retirement he took up painting and picture framing and in later life he would invite people into his house look at his pictures.

At the Heart of the Boat

John B. 'Johnny' Drew, Assistant and Station Mechanic 1934-1970

"The Coxswain is in charge. He has a great responsibility, especially in rough weather. He is in charge, but my argument is this: It doesn't matter how good the Coxswain is, if he hasn't got a good mechanic with him, he can't do a thing.**"**

However much the Coxswain has mastery of his craft and his crew, and takes overall responsibility for their operations, it is always the mechanic who spends most time with the vessel, usually on a daily basis. The station mechanic is employed full-time by the RNLI to keep the boat ready to burst instantaneously into action and, in addition to the engineering side of the job, he will also be the person to know most about the rest of the boat's condition. He will know every inch of the wooden hull from external inspection while lying beneath the keel, to internal knowledge acquired through hours of working with his head deep in the bilges. He will know the condition of the gearboxes, shafts, props, rudder and steering gear. He will be aware of the levels of fuel, lube oils and cooling water. He will have regularly inspected all the navigation and safety equipment: the pumps, alarms, electronics and communications devices, the radio and spotlight, as well as the skin fittings, the deck equipment such as windlass, anchors, warps and towing ropes, and the various cleats, bitts, fairleads, ventilators and bollards. And all the brass and copper work will have been lovingly polished and buffed to a high shine. In short it will be 'his boat' and it will be cared for with fierce pride. The mechanic accompanies the boat when it is taken out of service for maintenance or repair and, therefore, he would not be available to join his crew mates on the relief boat.

John Batten 'Johnny' Drew was born in 1905 into a fishing family in Mousehole, and he had first-hand memories of just about the whole of this story, having served for 29 years with all three of the *W&S's* Coxswains. In an extended interview, tape recorded in the 1980s, he recounted some of the most dramatic moments of his career in the context of a fisherman's life in this small maritime community, which epitomises the traditional image of a lifeboatman. Johnny recalled the opening of the Penlee station in 1913, when the open boat *Elizabeth and Blanche II* was transferred from Newlyn. He just missed out on being selected for that boat's final shout at Penlee, but he made his lifeboat debut as a volunteer crew member on *The Brothers* in 1923, at the age of 18. He transferred to the *W&S* on its arrival in 1931, and in 1934 he was taken on as assistant mechanic to Joseph Madron. He carried on his trade as fisherman, but four years later he was fortunate to get full-time employment on the lifeboat as station Mechanic, a position he held until compulsory retirement in 1970 at the age of 65. He had served on various Penlee lifeboats for a total of 47 years.

All those who knew Johnny remember him as a real character.

But he was also one of the best mechanics. Nim Bawden, his assistant mechanic for 10 years, recalls that "He was fanatical. After every launch the crew would wash the boat down with a hosepipe and we would chammy it off and then use old cedar polish with a dry cloth, and 'don't forget to clean behind the ears' — polish both sides of the propellers." The maritime artist Geoffrey Huband, who became good friends with Johnny while holidaying at Mousehole, remembers: "You could always tell what he was thinking. He was very fastidious." Johnny had even stripped the cowl ventilators of their dull paint to reveal the brass, which he polished diligently.

Johnny was also a bit of a prankster. Geoffrey Huband recalled that he would make a routine radio check every day to the neighbouring stations at The Lizard and Coverack. One day, for a laugh, he called The Lizard and told them some of the 'high ups' from London were coming down on an inspection and would visit their boathouse. When the fancy car arrived, the Lizard crew were amazed to see JD and some of his mates pile out. It had just been a joke; Johnny had hired the car and driven down to The Lizard to wind them up.

He was not much of a road user, however. He was once driven the nine miles from Mousehole to Sennen by Henry Nicholls, the Sennen Cove Cox'n, and that was the first time in his life that he had ever been there by road. He did eventually buy a car, a Ford Popular, but he couldn't really drive it. Nim recalls being driven down to Newlyn in it and when Johnny turned left to the slipway he hooted twice. "Why? well, we're turning to port." When pulling out in reverse he hooted three times.

Johnny was not a regular drinker, although on one occasion he was with an Australian lifeboat cox'n who got him drinking rum and shrub. And Nim recalls that day he did get drunk. "He came home ready to carve the chicken and the damn thing flew off the table, but he managed to trap it against the front door." After the death of his wife, Johnny took up smoking and always had a cigarette hanging out the corner of his mouth. Nim, who used to collect the fuel from the local garage, carrying 10 or 12 cans in a handcart, remembers seeing him pouring petrol into the tank with an unlit fag on.

Johnny Drew (right) and Eddie Madron, with whom he served for 32 years on three lifeboats
EDWINA REYNOLDS COLLECTION

The man with the spanner and the oil can

"These engines on the *W&S* were the RNLI's own design, built by the Weyburn Engineering Company. But the RNLI had their draftsmen and these engines were built to their own requirements. They had dual ignition, high tension mags, no coil, magneto. Electric hand starts and everything on them was duplicated. Two sets of plugs, two sets of mags, everything. The point was that they were in constant use. One wasn't for a standby; the two mags and the two sister plugs were in action all the time. It wasn't a question of switching from one to another; they were in constant use. I had them in such a state that I felt proud of them in the end. The speed of the *W&S* was 8 knots. Lifeboats, and their engines, are designed to maintain their maximum speed under the worst conditions for the safety of the boat. More or less. If some of these engines were put in ordinary boats they would get much more speed out of them.

In 1938 I was appointed from Assistant Mechanic to full-time Motor Mechanic. I had no previous mechanical training whatsoever, other than when I was in the fishing we had our own fishing boat with Kelvin engines in them. Old fishermen had a fair knowledge of the petrol/paraffin engines which were normally Kelvins. They were the main engine for economic reasons and simplicity for fishermen.

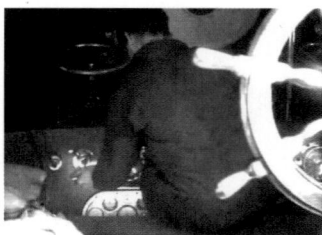

From 1924 that's when I reckon I started fishing in my own father's fishing boat the *Bonnie Lass*. The two engines she had in, one was a sleeve valve and the other was a pocket valve. Any cleaning or overhauling I would do all my own work on the engines. I stripped them, cleaned them, assembled them, I did everything. So I had a good general knowledge of the whole engine. I compliment myself that whatever was wanted I could do. And if it came to fitting a part that you ordered through the agency you would assemble the part yourself.

That stood me in good stead to be become lifeboat Mechanic. When I was taken on as Mechanic, I think the District Inspector didn't treat everybody like he treated me. I was on twelve months' probation. As I went on, I thought he was so strict with me at times that he was trying to catch me out. When it came to the dismantling, repairing everything — whereas the Mechanic used to have help supplied, such as a chap for labour — he wouldn't let anybody else help me, because I was on probation. Which I think afterwards was good, because I proved myself to him. He gave me a rough time but I think it proved

in the end and it gave me more confidence that I did the job and I proved myself. Towards the end he was the best one I could have had for Inspector. I suppose he proved me and I gave him every satisfaction, never let him down and therefore he rubbed it in on me, which I didn't appreciate at the time. It proved dividends in the end.

One day, the District Inspector called: "Can you be at the station 2 o'clock. I'm bringing somebody along to see you." About ten or quarter past two I could hear footsteps coming down over the steps. The door was opened and I could see him with two other chaps coming with him. They had lifeboat hats with propeller badges on them, which meant they were lifeboat mechanics, see. Of course, I had met them before, on occasion, perhaps through passing through on a boat, or somewhere else. Anyhow I said it was nice to see them. The Inspector said. "Now go on, it's up to you and Johnny now." I said "What is it boy?" And he said "We've come to see you start the engines." "To see me start the engines, why?" "Because we have two of the same type of engines in our boats." Entirely the same engines. So I said "That isn't any problem".

So I go aboard, unlock the engine room, open the hatch. Mustn't do a thing until he'd looked in and see what I do. I couldn't understand this. He said "I only want to know what you do. Because as you know we have the same type of engines as you. But the Inspector has told we how easy you start your engines and, many times, we have a job to start ours. I want to see what you do."

So when I opened the hatch and turned on the lights, he looked down and the two starting handles were in position for starting. Whereas they wouldn't normally be in the starting position, they would be in the rack — they could start by electric, you see. So I said "I always start the engines by hand every time I start them." "You do? I've got a job starting ours with electric." Anyway, now I had to do what the Inspector said, the normal procedure for me, what I always did. Pump up the air, turn on the fuel, flood the carburettors, with the handles being already shipped. The starting system with those engines was down to 3 to 1. Three turns of the handles to one of the engine. So every time, I counted. I took over the handle: one, two, three, she's gone. Started to turn the other handle: one, two, three, she was running. So the Inspector was looking and he said to them. "How about it now. You've been years you wouldn't believe me, would you. What about Johnny now?"

I had proved him right. He had brought two engineers that had more experience than me. But I had taken so much interest in the setting and having the right set up of the plugs and everything. I was very precise in that. I had it to a very fine art.

I forget what year it was, 1956 I suppose, we went for a refit at Mashford's Shipyard at Plymouth [actually at Cremyll on the Cornwall side of the Tamar] and we were away for six months. Of course, the mechanic went with the boat. He was in charge of his own machinery and working under the

Johnny started the engines with the crank handles to impress the visiting engineers but normally he used the electric starter

Cremyll boat yard was purchased by Messrs Watermans, late of Anderton, from Mr Ridley in about 1870. The yard was subsequently sold again to a Mr Rogers who in turn sold it to Messrs Mashford Brothers in 1937. The yard has a long and honourable history and its excellence of workmanship is recognised throughout the sailing world. It has been said that if a Captain of an RN Ship was presented with a Westcountry crew which contained shipwrights trained at Cremyll he went to sea a happy man. The yard is still functioning today, specialising in the maintenace of commercial craft and restoration of classic vessels.

regulations of the yard. Under the Institution, but under the regulations of the yard. There were five brothers, the Mashford brothers, and I can truly say they were all like gentlemen to me because I know I worked hard, but I enjoyed it. I had every facility put at my disposal and I took so much pride in them engines. The engines were completely taken out of the boat. I had the experience of seeing every nut and bolt of them two engines taken apart. Not just sections of them but everything was taken apart during the refit. Because she was away from the station for six months that gave me a good chance, not only on the mechanical side, but also all of the engine casings and that were aluminium and they were all painted white. I stripped all the casings down to the clean metal, not a speck on them, and built them all up with two or three coats of undercoat and gave them the final coat when they were assembled. There was hundreds of hexagon nuts from 5/8ths down to 7/16ths all done out in black on a white background. So you can imagine what they looked like. I shall always remember when they were all dismantled and the engines were in their trays with all the parts washed and ready for renewal and assembled, the Superintendent Engineer came to visit the yard and he said "There isn't much of it left is there? I'll come and see it when it is completed." Which he did after the six months. They were on the stands, the two of them, all built up, all repainted ready for starting with the handles in them, ready to be installed in the boat. He looked at me and said, "If they work as good as they look, you won't have many snags." And that's as it was. I never had a snag.

They were lovely engines they were. I would back my life on them. I suppose there wasn't hardly any lifeboat that spent so long on station as that *W&S* did. She was there 29 years. The normal life of a station lifeboat was about 20. Same engines from 1929 until 1960.**"** [And for another decade in the Reserve fleet.]

On his retirement in 1970 Johnny said to his successor:

"If you've got anything to do don't put it off till tomorrow because you might be called out tonight. I spent hours out in the lifeboat house in the evenings and night times. Nobody knew I was out there. My wife would ring me up, sometimes she would come out with me. If I had something to do, I did it. But it paid dividends, gave me peace of mind and the confidence I knew it was right.**"**

Those who served on the *W&S*

There were just two Honorary Secretaries during the boat's 29 years of service at Penlee: Barrie Bennetts (1919-1957) and his son J.K. Bennetts (1957-1964).

The three Coxswains were Frank Blewett, Edwin Madron and Jack Worth. Second Coxswains: Richard Johns, Edwin Madron, Joseph Madron, Jack Worth, John Wallis. Motor Mechanics: Joseph Madron and John Drew. Assistant Mechanics were Drew and Jack Wallis.

Many volunteers ran for the boat and not all of them succeeded in winning a place on the crew. A search through the Service records has come up with the following names which might not have appeared prominently in the ongoing narrative. Where possible we have given the date they were included in the crew list:

John Cotton 1932
William Harvey 1933
William Monkton 1935
Benjamin Jeffrey 1936
Arthur Johns
Henry Eddy (Harvey?)
William Quick
Norman Wallis
Edward Tregenza 1937
Benjamin Jeffrey
Richard Humphreys 1937
Luther Oliver 1937
Arnold Gartrell 1937
Bertie Jenkin 1938
R. W. Johns Jr
William Downing
N. Downing
John Maddern
George Torrie
Harry Blewett
Clarence Williams 1940 (radio op)
William Roberts 1940
Ben Pender 1941
James Pentreath 1941
Richard Sampson
Nicholas Richards
D. Blewett (1942-44)
A. Eddy 1949

Trevelyan Richards 1950
A.W. Matthews 1950
M. Torrie
J. Halse
W. Hoare
H. Bartlett
D. Cotton
C. Chigwin
Richard Richards
E.F. Madron Jr
J.J. Madron Jr
J.B. Madron
W. Downing 1951
B. Downing
T. Chivers 1952
J. Quick
M. J. Cobb
Clifton Pender
W. Pender 1959
Ned Tregenza
T. Tregenza
Owen Ladner
R. Pomeroy
G. Beare
H. Drew
D. Cornish
Nimrod Bawden 1960

The final crew of our boat, lined up for the dedication of its replacement (Left to right): Arnold Gartrell, Owen Ladner, Clifton Pender, Clarence Williams, Jack Worth, Jack Wallis, Johnny Drew and Ned Tregenza

A warm welcome at Penlee

DUDLEY PENROSE / EDWINA REYNOLDS COLLECTION

WITH flags and bunting streaming, the fishing craft in the harbour gaily decorated, and with a line of streamers from the bow of the new lifeboat to the masthead, Mousehole on Saturday, August 15th 1931 greeted the presentation of a new lifeboat to the Penzance-Penlee branch of the RNLI. Happily, the weather, though boisterous and threatening, held up during the ceremony, which was witnessed by a large concourse of spectators. [As recorded by *The Cornishman*]

The boat, which is a handsome craft, was described as 'the last word' in lifeboat construction. She was presented to the Institution by Miss Coode, [niece of one of the donors] in a graceful speech and has been provided by bequests of the late Miss Winifred Alice Coode and Miss Ellen Young. The Mayor of Penzance (Ald. Richard Hall) presided, and was supported on the platform by the Mayoress, Miss Coode, Mrs Favell (Penberth), Col. The Master of Sempill (*aka* Lord Sempill, a member of the Committee of Management of the Institution). Lieut. Com. H.L.Wheeler RN (District Inspector of Lifeboats), Rev. J.F. Prideaux, (Vicar of Paul), Rev.W. Rickard, Ald. Charles Tregenza J.P. (member of the Committee of the Penzance-Penlee branch), and other dignitaries, including Mr Barrie B. Bennetts (Hon. Sec. of the local branch).

Lieut. Com. Wheeler then gave a brief description of the lifeboat and concluded: "This boat is the last word in lifeboat construction. No money has been spared in getting everything that is of the best. She is capable of carrying 142 people and I can safely say that this station is now one of the finest equipped stations in the British Isles and, with her able crew, may be depended on to uphold the traditions of the service." [Applause]

Miss Coode, in a very graceful speech, said: "I am here today to name this lifeboat and present her to the Institution, and I am very honoured indeed, for I feel that the inauguration of a lifeboat is a very important event. The Lifeboat Institution cannot but be essential in a country that is an island [Hear!, Hear!] and the proof of that lies in the fact that the lifeboats were first invented and constructed in England, and nearly all the other countries that have lifeboat institutions have

founded them on the lines of our own [Applause]. We lead the way, for 'necessity is the mother of invention' and it is because of what makes this necessity that you now have this wonderful new vessel, which so far surpasses anything which could have been imagined in 1782, when the very first lifeboat was conceived and designed by Lionel Lukin, a London coach builder. It was he who sowed the seeds for the RNLI, which was founded under another name in 1824." The Rev. FJ Prideaux dedicated the lifeboat, and Miss Coode named her the *W and S*.

The 'Centenary' vellum, signed by HRH The Prince of Wales in May 1930, acknowledges the services of the Penlee (Penzance and Newlyn Lifeboat) which was actually established in 1803

Mrs Molyneux Favell presented Ald. Charles Tregenza with the centenary vellum awarded to the Penlee lifeboat station. Mrs Favell remarked she felt it an honour to do that.

The vellum had been signed by HRH The Prince of Wales as President of the RNLI. Penzance-Penlee has had a lifeboat for 100 years and during that time no fewer than 324 lives had been rescued [Applause]. The succeeding crews had always upheld the very highest traditions of the lifeboat service [Applause]. Ald. Tregenza said it was a great pleasure, on behalf of the Penzance-Penlee branch of the RNLI, to accept that vellum at the hands of Mrs Favell whom they were all delighted to see there. Mrs Favell had a very humane feeling for all on sea or land [Applause]. That vellum celebrated the centenary of the lifeboat Institution at Penzance during which time, as Mrs Favell had remarked, between 300 and 400 lives had been saved, and the crew was always ready to respond in rough weather. He would have liked for visitors from inland to have seen the former boat when she put to sea in 1929. She was caught by a wave which those onshore thought would have rolled her over and washed her onto the beach, but she proved a fine craft and by good seamanship she was steered through the huge waves .

Col. The Master of Sempill cordially congratulated Miss Coode for her very remarkable and splendid speech. "I thank you too for having invited me to be the means of representing the Lifeboat Institution and of passing this boat over to the local branch. . . As Miss Coode has told you, her aunt left a certain sum of money to the Institution many years ago and also Miss Ellen Young did the same thing. Unfortunately the sums left were not sufficient to purchase a boat of this description — as you know boats have become very expensive nowadays — but these two sums were kept for a period of time and eventually, when they were put together they were sufficient to purchase this boat for the Institution and for service on this station [Applause]. I should like to convey to the committee of management of this branch and to the Coxswain and crew the very good wishes and congratulations of HRH The Prince of Wales 'our President' and of the members of the management committee for the splendid services which have been rendered and are being rendered here."

Eagerly jumping aboard

'Huge creamy crested billows were hurling themselves shorewards; the wind howled dismally; storm and rain clouds lowered overhead. It was under these conditions, as reported elsewhere, that the dedication of the new Penlee lifeboat took place at Mousehole, on Saturday. It was after the pomp of ceremony that the fun started — if tossing on the foaming sea can by any stretch of imagination be called fun. It was decided to take those who were venturesome enough for a trip on the new craft. A *Cornish Evening Tidings* reporter stood on the quay; he looked at the sky, then at the sea, and his heart sank. But all of a sudden the journalist in him became predominant, and he murmured. "Well, there may be copy in this," and straight away he jumped aboard. "By this time," he writes, "small boys and adventurous Miss 1931s were eagerly jumping aboard. Little did they reckon what was in store for them. There was much excited chatter: facetious remarks were being made all over the boat: everyone was asking "How are you feeling? Are you feeling sick?" But the effects of *mal-de-mer* were not making themselves felt as yet.

Rain was falling, everything looked drab and dreary when eventually we cast off, as happy and excited a crew as ever a lifeboat held. We had just moved out of the harbour when the storm started to make itself felt. Quite enjoying myself, I had my head just above the for'ard deck, when we shipped a brute of a sea and I got it in the neck: that is to say full in the face, and became uncomfortably wet before the voyage had started. Down below we all got, one Miss 1931 had got the full force of the wave and crawled down beside me, drenched. The voyage out had started: we were packed down below like a rugby scrum. Up and down she went: this moment poised on the top of a wave, the next we felt as if we were being lunged to abysmal depths. Then the sea would strike us and the water would come pouring in. Assistant Coxswain Dick Johns called out "This is nothing. This is fine weather." Fine weather! I ask you! But in spite of the heavy sea no-one seemed at all distressed, and everyone was shouting and, apparently having a most jolly time. She turned at last, when I should imagine from my position beneath the deck, she got opposite the Coastguard's look-out, and the run with the wind and tide was not so exciting.

The heat below was becoming so oppressive that I was glad to climb on the deck again, although it was still raining. There was St Clement's Isle wreathed in foam, and I thought, as I saw Coxswain Frank Blewett unerringly steer the craft through the seas, that we, the landlubbers, do not realise what it is like to go to the rescue on a dark winter's night, when 'wind's like a whetted knife'; and we can never realise the peril these brave men run on their errands of mercy. It is in infinitely worse weather they venture forth; and we must admit that the assistant Coxswain was right when he remarked "This is fine weather". I remarked to a friend of mine when we came ashore that the trip had been far enough for me. He agreed.'

Assistant Coxswain Richard Wallis 'Dick' Johns Snr (above) and Coxswain Frank Blewett enjoyed the 'fine weather'

A Life of Service

**The boat's first call to action came just short of
three weeks after the dedication ceremony**

THE *W&S* (ON736) is credited with saving 102 lives during her time 1931 SEPTEMBER at Penlee, but her first service, which took place just a few weeks after the welcome ceremony, was the opposite of a triumphant rescue. Instead of celebrating lives saved, the crew returned to the boathouse with heavy hearts, bearing a body.

The corpse landed on **September 4, 1931**, was that of Captain John Campbell, Master of the steamship *Opal* (573 tons) registered at Glasgow, and bound from Antwerp to Cardiff with a cargo of grain and maize, which foundered in heavy seas off Land's End. The cargo had shifted, creating a dangerous list which would prove to be catastrophic.

As reported in the *Western Morning News*, the ship was first spotted by visitors staying at the Land's End Hotel when it was about three miles SW of the Longships lighthouse, which is only about a mile and a quarter from Land's End.

'Looking out to sea at about 4.30 pm they dimly perceived through the thick mist a small coasting steamer battling her way through mountainous seas. There was terrific NW gale blowing and heavy waves could be seen crashing over the unfortunate vessel as she drifted helplessly on, unable to make an effort to reach a safe port. She was on her beam end, and it was evident to the watchers on the cliffs that her position was perilous in the extreme.

'The nearest lifeboat was at Sennen Cove and they were informed of what had been seen, but the weather conditions were so bad that the

*The first shout
for the W&S
was to a serious
incident, looking
for survivors of
a shipwreck off
Land's End*

EDWINA REYNOLDS
COLLECTION

N. & S.

Royal National Life=Boat Institution.

LIFE-BOAT ___W and S___ stationed at ___Penlee___

RETURN OF SERVICE on the __4th__ day of __Sept__ 19 31

Date and Circumstances of the Case.

About 4.50 p.m. on Sept 4th the Coxswain received the attached message stating that a
Ship was in distress 3 miles S.W. of Lizard. The Coxswain at once called out the crew & got the boat afloat. The
crew were all assembled & the Boat left the Slipway at 5.15 p.m. The lifeboat arrived on the
vicinity of casualty about 6.15 p.m. When a ship lifeboat was sighted. The lifeboat (came
to waterlogged and with no one aboard. A search was made amongst the wreckage
for survivors & the body which proved to be that of the Captain was found. There
was no trace of any other. The Captain's body was taken onto the lifeboat
for safety. The search was continued until the bay proved empty. As there was
another to run to be in case men were drifting in boats. After searching about
10 miles to leeward finding nothing, it being dark, it was decided to make for Newlyn.
The boat arrived there at about 10.15 p.m. The body of the Captain was handed onto the
Police. She had remained in Newlyn Harbour for the night owing to weather & could not be known
that night.

QUESTIONS. | **ANSWERS.**

(a) Bearing and Distance of Wreck? ... (a)

(b) Direction and Force of Wind? ... (b)

(c) Condition of Sea on Sands? State whether "smooth," "moderate," or "heavy." ... (c)

(d) State of Tide? ... (d)

(e) Signal fired? ... (e)

1. Rig, Name and Port of vessel? ... 1. S S Oper . Glasgow

2. Names of Master and Owner? ... 2. Capt John Campbell ? M Ln Robertson

information from Hasborough or Cross Sand Light Vessels.

lifeboat crew, despite their strenuous efforts, could not get their boat to sea. A decision was then made to call out the Penlee lifeboat which is situated about seven miles along the coast, and at about 5.20 the new Penlee motor lifeboat, manned by the men of Mousehole, put to sea on her first errand of rescue. She had to combat a terrific headwind, and there was a very strong sea against her.

'She found great difficulty in rounding Land's End, and before she could arrive at the scene the ill-fated vessel had sunk beneath the waves before the eyes of those on the cliff anxiously watching the drama which was taking place. The lifeboat made an exhaustive search in the vicinity of the Longships in the hope of picking up any survivors who may have been clinging to pieces of wreckage. Although there were several steamers passing in the vicinity none of them made any effort to search as they were apparently unaware of the catastrophe.'

That claim in the news article was later disproved, as at least three ships were involved in reporting the sinking, and standing by or rescuing the surviving crew. One of vessels that first became aware of the event unfolding was the fisheries protection gunboat HMS *Spey*, which picked up several lifebelts belonging to the *Opal*, and sent a wireless message at 3.32pm: 'There is a ship in distress off the Longships. I am unable to assist.' Another message followed an a hour later: 'Steamer foundered at 4.30 GMT (5.30 Summer Time) 3 miles South-West Longships Lighthouse'. A German steamer, the *Max Albrecht*, also radioed: 'Three ships in position 3 miles SW Longships looking for survivors.' She reported passing four floating lifebuoys marked SS *Opal*, Glasgow, and one boat half-filled with water, with no one on board. That vessel and two other ships were reportedly standing by, one of which, the *Wild Rose*, of Liverpool had rescued nine crewmen and one passenger.

A typical ship of the period rounding Land's End and heading into the English Channel, with Longships lighthoue in the distance

It transpired that the crew of the *Opal* had been ordered to abandon ship and take to the open boats by Captain Campbell, one and a half hours after the cargo had dramatically shifted. Captain Campbell, who was washed off the bridge, and the Chief Engineer William Daly, both from Glasgow, were the only crew members not to be saved. Both were well known at Newlyn, where the vessel had frequently docked.

The *Wild Rose*, which was also having a difficult time rounding Land's End, saw the distressed seamen in their boats and immediately went to the rescue. The sailors, who had undergone much hardship and were suffering from exposure, signalled to the *Wild Rose*, which fortunately spotted them and drew alongside. One of the lifeboats containing six men had been upset three times, and the crew had been fortunate to stay afloat by hanging on to barrels and pieces of broken wreckage. Lifelines were thrown to the crew and after much difficulty they were got on board. Both Penlee and Sennen Cove lifeboats launched and made for the area, unaware that 10 survivors had been rescued by the *Wild Rose*. The W&S recovered the Captain's body, and the *Opal* herself foundered three miles South West of the Longships.

The *Opal* was owned by J. Wm Robertson of Glasgow, her port of registry Glasgow. She was 573 gross tons, built and engined in 1919 by Messrs J Lewis and Sons, Ltd, Aberdeen, and originally named *River Dee*. The firm owned 38 vessels, all under 1,000 tons

A member of the crew told the *Cornish Evening Tidings*: "About three o'clock on Friday afternoon, when we were abreast of Pendeen, the list started. It was due to the cargo shifting, and we turned back with the intention of sheltering in Mount's Bay. When got about three miles Southwest of the Longships Lighthouse, it became obvious that we could do no more, and we decided to abandon the ship. She was on her beam ends and gradually sinking lower and lower into the water. The lifeboats were got out with the utmost difficulty. We were six of us in one of the boats, and this boat capsized three times. We were thrown struggling into the water. We clung to pieces of floating wreckage, barrels, tanks, and the bottom of the submerged boat until we righted her and clambered aboard. This had to be done each time, and the long exposure in the icy

"We were six of us in one of the boats and this boat capsized three times

water was having a weakening effect on us. I never thought I should get through alive."

Another survivor, Second Engineer Matthew Martin, told the *Western Morning News*: "I was alone in my boat, and therefore I could do nothing to help those who were in the water, because it would have been futile for me to have pulled against such waves. I could see three men drifting in the water and recognised the three firemen, McFaull and the two Macauleys. The Macauleys were father and son. One was clinging to a barrel; another hung on to the meat safe, while the other secured a plank. The six men in the other boat could do nothing out of their plight, as they were in difficulties themselves. The [ship's] lifeboat was continually capsizing, they lost all their oars, and the ship was full of water. The SS *Wild Rose* then came up and stood by. First of all she took up the men in the water, by means of ropes. The sea was too rough for her to launch her boats with any degree of safety, and it was wonderful how she managed to get us aboard as it was." Mr Martin, as soon as he landed at Newlyn, rushed to his wife at Penzance, who was naturally greatly surprised and delighted to see him. He was deeply moved when he heard that his skipper was dead when picked up.

In another interview with the *Evening Tidings,* Mr Martin gave a very graphic account of his experience: "It was about 12.30 on Tuesday night when we left Antwerp, and we had a very dirty run all the way down the channel. I was second engineer on the *Opal*, and had been with her four years this month. She is owned by Messrs. W. Robertson, of Glasgow. She was built in 1919, and was a good, seaworthy ship. We had on board a cargo of maize for Cardiff, and it was packed loose in the hold, except round the bulwarks, where the grain was in bags. We were due to arrive in Cardiff at about noon on Friday, but the weather was so bad that we were only off the Land's End, and then the cargo shifted. The loose maize was the first to move, and the ship took list. We put about to run for safety, but it was apparent that the end was near when the bags around the bulwarks burst and the maize came out. I was down below in the engine room, with the Chief Engineer, most of the time, trying to get up steam, but it was impossible because of the considerable list. If we could have got up steam we might have had a chance, but try hard as we could, it was impossible.

"It was decided to abandon ship, but this was a hard task owing to the terrific seas. There were originally three lifeboats on the ship, a dinghy and two larger boats. The dinghy was smashed by a huge wave crashing on her. The other two boats were got out with difficulty, and six men scrambled into one of them. The Chief Engineer and I got out the

The *Wild Rose* was registered at Liverpool and owned by Messrs R Hughes and Co, of Liverpool. She traded regularly in these waters, often carrying copper from the mines at Amlwch in Anglesey

HUGHES COLLECTION

other one. The boat would have been washed away from the *Opal* were it not for the painter caught in something and held. I was able to jump in and shouted to the Chief Engineer to follow suit. He was standing only a few yards from me. He was wearing a lifebelt and was gripping the rigging. The boat was in danger of jamming against one the ventilators of the *Opal*. I ran forward to stop her, and at the same time I again shouted to him to jump. Suddenly a monster wave lifted my boat and swept her away from the doomed vessel, and that was the last I saw of Chief Engineer Daly. The last I saw of the skipper was when he was standing on the bridge, putting on a lifebelt."

Another of the crew told the Press. "Just before we left Antwerp, I was in a cafe and there was a pack of cards on the table. A girl came up and offered to tell my fortune. She took up the pack of cards and told me my fortune. Amongst the things she said was: 'You won't be in your ship very long.' and it seems that she was right too. I never thought I should be rescued," he added, "and I couldn't sleep last night for the thoughts of the evening's events. "

The lifeboat crew comprised Coxswain Frank Blewett, Richard W. Johns, Richard B. Richards, Joseph Madron (mech), Richard Johns Jr, John B. Drew, Edward Downing, Benjamin Jefffery.

Cox'n Blewett told a reporter he was well pleased with the new lifeboat: "She rode the waves last night like a duck."

The wreck of the Opal lies at 50.01.30N, 05.42.30W

1931 SEPTEMBER Less than three weeks after the first shout for the *W&S*, the St Ives fishing vessel *Iverna* struck a rock off Porthcurno, while engaged in line fishing in fog on **September 17**, and was stuck fast. As the boat began to sink, the *WMN* reported that the crew took to their punt and made for shore at Porthcurno. They stayed the night at Treen and the following morning were driven to Newlyn to retrieve their belongings. The Penlee lifeboat *W&S* had been launched at 4.40am and proceeded to the location. The boat went as near as possible to the shore but found that the crew had left. The lifeboat returned to station at 7.25 and was rehoused at once *The Iverna sank at 50.02.24N, 05.38.50W*

1931 NOVEMBER The last wreck of a commercial sailing vessel in Mount's Bay was recorded on **November 3**, when the French two-masted schooner, the *Sainte Ann*, carrying a cargo of coal, was driven ashore at Loe Bar, Porthleven in a SW gale. On the fourth day out on passage from Port Talbot to Vannes in France, the 1914-built, 185 gross ton vessel was caught in a gale off Land's End, sustaining damage to her rudder, and *The last sail* sprang a leak which necessitated the crew pumping continuously. She *trading vessel* had also lost some of her sails. She was driven into Mount's Bay and *to end up on* forced on the lee shore. Following a message from the Coastguard to the *Loe Bar was* Motor Mechanic at 1.20pm, the Cox'n was informed that a vessel was *the Sainte Ann,* apparently in distress 3 miles SW of Porthleven heading for Penzance. *whose crew* The crew was summoned and the boat left the slipway at about 1.40pm *were rescued by* and went straight for the position given, keeping near to the shore on *the LSA. Photos* east side of Mount's Bay in case she was drifting towards the shore — *show the vessel* proceeded as far as Porthleven but never saw the vessel. The *Shipwreck* *started to break* *Index* records that at 2.30pm the *Sainte Ann* was driven onto the sandy *up quickly over* beach. She struck with her bow, but was immediately swept broadside *night* on and rollers began to pound against her side. The Porthleven Life

Saving Association rescued the crew of six. It was low tide when the vessel struck, and it was feared the battering she would receive when the tide rose would break her up completely during the night. She was swept 100 yards up the beach by terrific seas during the night, and the following morning was lying at the foot of the cliffs with her side stove in and her back broken. All hope of saving her was abandoned. *Her final position was 50.04N, 05.18W*

The boat's only service recorded in 1932 was on **August 6**, when the Norwegian steamer *Sneffeld* of Bergen went ashore at Logan Rock in thick fog, along with a French trawler. The lifeboat was called out by the Coastguard and arrived on scene around 4pm, to find the ship had refloated on the flood tide and proceeded on her course. However, a man in a punt told the Superintendent Cox'n that another steamer was ashore about a quarter of a mile further westward. The lifeboat found the large French steam trawler *Ancient Marie* on a ledge. The Cox'n spoke to the vessel but received no answer to his questions. The ship was then refloating on the tide, and when afloat she went seaward in the fog. The lifeboat returned to station 'ready for any other casualty'.

On **March 11, 1933,** there was a report to the Penlee station that an unnamed Italian steamer was in difficulties, but the lifeboat was not launched.

A 'Mayday' call was sent out on **August 4** by the steamer *David Dawson* of Newcastle from the Runnel Stone Rocks. Supt Cox'n Blewett recorded that when he arrived at the boathouse he learned that the steamer did not need immediate assistance 'But I thought it advisable to go afloat in case she slipped off and foundered in deep water. I left the slipway about 11.40am and reached the steamer about 12.30pm. I found she had got away alright and returned to station and housed the boat at 2.pm'. Frank Blewett added in his report: 'In view of the

The David Dawson sailing under the name of Koolanga some 10 years before the incident

fact that both the Sennen and the Lizard lifeboats were off service, and the journey from Penlee to the Runnel Stone Rocks takes about one hour, it was thought advisable in spite of the last message from the Coastguards to launch the boat and proceed to the ship.'

The *David Dawson* (5286grt) was built in 1918 as the *Sutherland* by W. Doxford, was renamed *Southmead* in 1920, *Koolonga* in 1921, *Caithness* in 1929, and *David Dawson* in 1935, owned by Jubilee S.N.Co, Newcastle. In 1936, now known as *Avon River*, of Bristol, she was wrecked in Hudson Bay. The vessel was recovered later and sailed again as the *Sutherland*.

There are no records of any services during 1934, although the *Cornish* *Telegraph* included a couple of references to the Penlee lifeboat.

On **May 31**, the paper reported that a cinema film was being shot in Mousehole that included sequences with the lifeboat. Originally

titled *The Lady of Pendower*, this 58-minute drama was eventually released in 1935 as *Breakers Ahead!*. Written and directed by Anthony Gilkison, the film starred Barry Livesey, April Vivian and Roddy Hughes. The story was briefly outlined as 'Fisherman tries to drown rival, and then dies to save him from storm'. (Not to be confused with the 1938 release of the same name, but different story, which was filmed in Polkerris.) The skipper was played by Richard Worth, a cousin of lifeboat crew member and future Cox'n Jack Worth.

The paper reported on **August 2** that the Penlee lifeboat was back on station after a refit at Falmouth.

1935 NOVEMBER On **November 30, 1935** the *Cornish Telegraph* reported that soon after midnight a fishing boat was seen burning flares South of Porthleven. There was a strong WNW breeze, rough sea, heavy rain showers. 'The lifeboat launched at 12.45am and found MFB *Adventure* (PZ141) of Penzance, with a crew of five, riding at anchor close to a lee shore near Mullion Island. Her engine had failed and she was in danger of being driven ashore as wind and sea were increasing. Lb towed her to Newlyn harbour and returned to station at 5.15am. Saved vessel and five persons.' Cox'n reported 'drogue rope broken'. The owner and master of the *Adventure* was Richard Richards, a lifeboat crew member who was on the crew during the boat's previous service to the *David Dawson*.

1935 DECEMBER At about 5am on **December 8**, a message was received that a vessel was firing distress signals in Mount's Bay. The *W&S* found the steamer *Cornish Rose* at anchor with a crew of nine on board. The Cox'n spoke to the Captain who informed him that the ship was leaking badly. The Cox'n asked him if he proposed abandoning ship and he replied 'No', but he requested the Cox'n to get him into Newlyn harbour. The ship had no steam and its fires were out. The lifeboat towed the steamer into harbour arriving there at 7.30am.

On December 6 the SS *Queen City* had reported a small vessel firing distress rockets 24 miles NNW of Pendeen on the north coast. The ship was taken in tow by the SS *Finland* off the Runnelstone buoy at 1pm on the 7th. By 8.10pm she had towed the *Cornish Rose* into Mount's Bay where wtaer started to rise in the engine room. The *Western Morning News* gave further details: 'The *Cornish Rose* had developed engine trouble in a strong northerly wind, and although she was picked up very soon after by the *Finland*, heavy weather made towing very slow. The tow rope parted three times. On reaching Mount's Bay the *Cornish Rose* was docked at Newlyn to await repairs. She is in the charge of Capt Owen.

'The *Cornish Rose* was bound from Fowey to Preston with a cargo of china clay [. . .] Capt Owen said that when they left Fowey on Friday

there was a strong northerly wind and a heavy sea. At about 6pm, when they were 14 miles North-North-West of Pendeen, they developed engine trouble. "We signalled for assistance and very soon afterwards the *Finland* picked us up. By 9 o'clock she had us in tow and we proceeded to Mount's Bay, which was the nearest refuge. Owing to the weather we could proceed at only a slow pace. The towing warps parted three times, and this entailed a good deal of manoeuvring to get us in tow again. We reached Mount's Bay about 5.30pm on Saturday, but the *Finland* was too large a vessel to bring us into harbour so we dropped anchor in the bay until 5 o'clock on Sunday morning, when the Penlee Lifeboat came out and towed us into harbour." Capt Owen added that no one on board sustained injury, and he hoped to be on his way again in a day or two.'

A Bronze Medal service at Penlee

The *Life-Boat* Journal reported that 'A message was received at the Penlee lifeboat station at 2.15 in the morning of **January 27, 1936** from the Coastguard that a ship appeared to be on fire near Gear Rock in Mount's Bay. Twenty minutes later the motor lifeboat *W&S* was on her way. A strong SSW gale was blowing, with a heavy sea running, and the **1936 JANUARY**

MORRAB LIBRARY/ COLLINS

weather was thick with rain. In about half an hour the lifeboat reached the Gear Rock. There, by the light of her searchlight, she saw the 400-ton steamer *Taycraig*. The after half of the ship was submerged, and the crew of nine men were packed together on the forecastle head. The steamer was not on fire. In the heavy swell she had struck the Gear Rock while making for moorings and lay with her stern wedged on the rock, and the seas breaking over her. The fire which the coastguard had seen was mattresses burning as a signal of distress. Twenty minutes after making their signals, the men waiting on the forecastle head had seen the maroons fired at Mousehole, nearly two miles away. They knew that the lifeboat had been called out to their help, and were surprised that she came so quickly.

The *Taycraig* lay end on to the gale, so that there was no lee to give the lifeboat any shelter as she came alongside, and there was very little room for manoeuvring among the rocks. The Coxswain, however, succeeded in bringing her close to the starboard side of the forecastle, and threw a grappling-iron on board. The lifeboat was flung violently about on the big rise and fall of the seas, and every moment the captain of the *Taycraig* expected to see her come right on board the steamer, but the Coxswain skilfully kept her off. She touched the steamer, but was only slightly damaged.

Each in turn, and each watching carefully for his opportunity, the nine men of the *Taycraig* jumped on board the tossing lifeboat. Seven landed without mishap. One dropped right on the neck of the bowman. Another, misjudging his jump, fell between the steamer and the lifeboat, but he was seized at once by two lifeboatmen and dragged on board. The lifeboat then took the nine rescued men to Penzance harbour. She arrived at 3.25 in the morning. The service had taken just 50 minutes. It was a service in which the lifeboat was handled very skilfully and boldly, all the more so that the whole service was carried out with only one of the two 40 h.p. engines working.'

The Hon. Sec. Barrie Bennetts added the following remarks on the Return of Service Report: 'In view of the strong weather and the fact that only the starboard engine was working, this appears to have been carried out with singular ability. Both port and starboard engines were tested on Saturday and Sunday last and worked satisfactorily. One of the *Taycraig*'s crew, in jumping into the lifeboat, fell on the neck of the lifeboat bowman, Luther Oliver, and caused him (Oliver) injury. I have asked him to see his doctor and let me have the report.'

The Supt Cox'n reported 'The boat behaved well, in spite of the fact that the port engine could not be started.' [This was before Johnny Drew became station Mechanic.] There was no damage to the boat, 'but chain

Superintendent Coxswain Frank Blewett was the first of two coxswains of the *W&S* to be recognised for his lifesaving skills. His crew on the *Taycraig* shout comprised Luther Oliver, Jospeh Madron, John B. Drew, Bertie Jenkin, Benjamin Jeffery, Edward Tregenza and John Wallis. There were no medals for them but they did receive a useful bonus.

Patrons, Their Majesties The King & Queen

ROYAL NATIONAL LIFE BOAT INSTITUTION

FOR THE

Preservation of Life from Shipwreck.

(INCORPORATED BY ROYAL CHARTER.)

ESTABLISHED 1824.

SUPPORTED BY VOLUNTARY CONTRIBUTIONS.

President

His Royal Highness the Duke of York, K.G.

Chairman

Sir Godfrey Baring, Bt.

Deputy Chairman

The Hon: George Colville.

At a Meeting of the Committee of Management of the Royal National Life Boat Institution for the Preservation of Life from Shipwreck held at their Offices, London, on the 13th day of February, 1936 the following Minute was ordered to be recorded on the Books of the Society.

That the Bronze Medal of the Royal National Life-boat Institution be awarded to

Frank Blewett,

Coxswain of the Penlee motor life-boat, in recognition of his skill and seamanship, when the life-boat under his command rescued the crew of nine of the S.S. "Taycraig," of London, which was wrecked on the Gear Rock, Mounts Bay, in a strong S.W. gale with a heavy sea in the darkness of the early hours of the morning of the 27th January, 1936.

MORRAB LIBRARY/ COLLINS

plate on tipping table broken'.

The Branch Committee appreciated the 'smart, seamanlike and skilful manner in which the service was carried out', and to Coxswain Frank Blewett the Institution has awarded its Bronze medal with a copy of the vote inscribed on vellum and framed. To the Coxswain and each member of the crew it has made an increased money award of £2 17s. 6d. The total awards amounted to £27 14s. 6d.

During the following days the *Taycraig* received a severe buffeting which swept away her funnel. By February 10, she had apparently broken up, since parts of her were flung upon the promenade by heavy seas. *The vessel sank at 50.06.30N, 05.31.33W*

One of the *Taycraig*'s crew, Frank Gilbert (left), was born in Hayle, in Cornwall in 1914, and was a proud Cornishman. However, according to his son Andy, Newhaven became his adopted home after meeting and marrying Ellen Bryce. He went to sea at an early age and worked his way up from deckhand to an AB, then up through all the ranks of Bosun, Mate and finally Master. His time at sea was spent on coastal ships, cargo vessels, tankers and tugs. He was shipwrecked on more than one occasion, and at the age of 21 being rescued from the wreck of the coaster off Penzance, through an amazing display of skill and bravery by the Coxswain and crew of the Penlee Lifeboat, left him with an enduring respect for the RNLI. During WW2, he was serving with the AFS in

The *Taycraig*, seen here at Penzance, built in 1901 at Maryport, was of 407 grt, owned since 1929 by County of Cornwall Shipping Co Ltd, of Redruth. the eighth owners of the vessel, which was in ballast from Plymouth to Newlyn, only to be wrecked so close to its destination

Newhaven and, on one winter's night, he jumped into the icy waters off the North Quay to rescue a drowning soldier. For this he was awarded the Bronze Medal of the Royal Humane Society. He died in 2004.

A message was received on **June 19** that a boat had gone ashore near Carn-dhu. It was the motor fishing boat *Gleaner* of Penzance, with a crew of five on board. Cox'n Blewett reported: 'About 11pm I called out the crew and proceeded to the position given. The fog was very dense and I was obliged to go under reduced speed. I arrived in the vicinity of the casualty and after searching about I found a fishing boat which had been ashore and had now refloated. I spoke to crew of same and they said their craft was making water and asked me to stand by them while they were proceeding to Newlyn harbour. This I did and found the pumps could keep her afloat until she arrived at Newlyn. I then returned to station and housed the boat around 6am, June 20th.'

On **December 21**, the SS *Mina* of Parnu (Estonia) broke her rudder shaft in a strong SW breeze and rough sea. The RMS *Scillonian* saw her and stood by, going alongside the casualty when about one mile from Wolf Rock. The Sennen Cove lb was launched at 11.50am and the *W&S* from Penlee at noon. The Sennen boat reached the casualty when about 8 miles SW of Tol-Pedn look-out. The *Scillonian* had by then got a line on board and was trying to tow her. The lb passed another rope between the steamers, and later put one of her own on board to help with the tow. Shortly afterwards the *Mina* sheered badly and the lifeboat had to drop her rope. In a short while the other two ropes also parted, but by then the *Mina* had made temporary repairs and was able to go her own way. When the Penlee lb arrived Cox'n Blewtt spoke to the captain of the steamer who said he did not require any assistance.The Sennen cox'n knew he could not rehouse in the weather conditions at Sennen so ran for Penzance. The *W&S*, also kept from rehousing by the weather, spent the night at Newlyn. Tugs arrived later tow the *Mina* to Falmouth .

Almost one year after the *Taycraig* sinking, another major incident, this time with a tragic outcome, occurred on **January 11, 1937**, when the Belgian MFV *Vierge Marie* (O-332) went ashore near Newlyn. At 7.40am, local police reported a wreck under Tregiffian Cliffs, near Tater Du Point. There was a strong SSW breeze blowing, heavy sea and thick foggy weather. The *W&S* launched from Penlee Point at 8.00am, steamed down the western shore, and found the motor trawler *Vierge Marie* of Ostend ashore and being pounded heavily by the rough seas. Four men had been attempting to launch the small boat when a huge sea came on board and washed them over the side. The lifeboat crew located three men in the water and hauled them on board. They were each given mouth to mouth

The *Vierge Marie* was a 26 metre steel trawler of 112grt, built in 1931. Powered by a 47 hp diesel. It was ownd by Emile Lus of Ostend

resuscitation — sadly, only one of the men was revived, and he died later after being landed in Newlyn. The fourth crew member also drowned.

The trawler had been heading for Newlyn from the fishing grounds with a crew of six. With no sign of the remaining crew being found, the *W&S* returned to Newlyn, landed the bodies of the three fishermen, and returned to station at 9.30am. The lost crew were seamen V.H. Maertens, E.J.

Dewaele, A.E. Easton and engineer A. A. Huisseune. The other two crew had managed to scramble ashore.

     Lloyd's List reported from the Coastguard at Penzer Point: 'Ashore at Treveen. Vessel being pounded by heavy seas, stern broken away. Skipper Emil Lus and a deckhand were saved by a rocket apparatus crew when they were trying to climb the cliffs. The skipper later said that the trawler had developed engine problems off Land's End and they could not avoid going ashore. By January 12 she was completely broken up by heavy seas, only her keel plate remaining.'

The vessel was lost at 50.05.51N, 05.32.20W

Mechanic did his best — 'couldn't do more'

The assistant mechanic, Johnny Drew recalled his experience of the event.

“On this particular morning, about seven to 7.30am on January 11, 1937, Bang! Up goes the maroons. Lifeboat! Well, you jumped out of bed, wasn't half dressed. Ran down, couldn't get out of the house quick enough. On this occasion I opened the front door and there was a motorbike coming down. Well, being a youngster I jumped on that quick enough. I was the first out there. Couldn't understand it. I thought 'where's everybody to, maroons have gone'. A minute or two later the Coxswain and Boat Launcher arrived. Anyhow, went into the boathouse and started the engines ready. Eventually we launched. It was a trawler ashore near where Tater Du light is now, on the Tregidda Point. The *Vierge Marie* it was called and as we went along the shore and up past Lamorna we could see these people waving on top of the cliffs, it was daylight now. As we were closing in, it's something I would never forget, but you can always repeat the truth: It happened.

     We could see the trawler ashore and we could see men in the water. And some of our lifeboat crew, one or two of them, did what they shouldn't have done; they took a line and jumped overboard to get to the men. But the next thing I was looking around and I was in charge of the lifeboat and the engines. Nobody to give me instructions. The Cox'n and the Mechanic had jumped overboard too and they couldn't do nothing. So, therefore, there was chaos. Four or five of the trawler's crew we got aboard the lifeboat. On the way back to Newlyn, I was having to do artificial respiration on them, but they all died. That was a very nasty sight and they were all young men. Awful to see them up in front of you, they were somebody's loved ones. That was a very sad occasion for peace time. There was an inquiry about it but I don't know the proceedings because I wasn't called. I did my duty but there was some inquiry concerning why this man and that man left the boat. Whether it was for the best or the worst I don't know.

It wasn't a very nice experience to be pumping these men on artificial respiration and the boat rocking. It was 'one, two, three, pump' until you revived the man. On three occasions that I experienced it I always said 'how can you save a man's life like this', because he was lying down and you would have your legs across his back. Now my contention it was then, and it always bothered me, with the lifeboat rocking how could you put the pressure on that man with the boat rocking. It was almost impossible. Well the answer was 'do your best'. You couldn't do more than that.**"**

On **March 25** Cox'n Frank Blewett wrote in the Returns of Services **1937 MARCH** book: 'About 8.30pm I received a message from Coastguards at Penzer Point Lookout which stated that the fishing boat *Betty*, of Inverness, was reported South West by West of Lamorna with engine broken down and requiring assistance. I considered the position given and asked Coastguards to report to their District Officer to inform Scillies. Soon afterwards I again heard from CG that this had been done and also Land's End Wireless station had been informed and broadcast to all ships to look out for the *Betty*. I now got in touch with our Hon. Sec. and reported all news to him. Shortly after, I received a message from Customs Officer Newlyn stating *Betty* was seen making for Newlyn with wind SW. I found out that this was about midday. By now a Northerly gale was blowing and it was feared the *Betty* might be blown out to sea, so we decided to launch the lifeboat and search for her. The drifter *Efficient* also went in search of her and, after an all-night search and failing to locate the *Betty*, we returned to station [at 9.30am].' The vessel was towed in by another fishing boat *Asthose*. The sea conditions were described as 'rough' and the weather as 'bright at times with hail showers'.

At 2.55am on **October 30,** the Penzance Coastguard telephoned that **1937 OCTOBER** the small motor boat *Apapa*, with one man on board, was missing from Newlyn. A NW breeze was blowing, with a moderate sea and rain. The motor lifeboat *W&S* was launched at 3.20am, and found the boat about three miles from Mousehole. She was at anchor, and the man was exhausted. He was taken on board the lifeboat and given stimulants, and his boat was towed into Newlyn. The lifeboat returned to station at 4.45 am. Rewards, £16 14s.6d. The Hon. Sec. noted that 'In my conversation with the Cox'n he informed me that there were no other craft available and I thought it advisable to use the lifeboat'.

During the night of **April 22, 1938**, the Coastguard at Penzer Point **1938 APRIL** telephoned that a fishing boat was in distress about two miles SSW of the point. It was the steam fishing boat *Pioneer* (PZ277) of Penzance. About 10.22pm, Cox'n Blewett reported: "I looked out from my house and saw the craft continuously showing flares and every indication of something being wrong. I at once summoned crew, launched lifeboat, proceeded to the craft and found her with her anchor down. I asked the skipper what was wrong and he said he had lost his propeller. I took him in tow to Newlyn harbour and then returned to station. The weather being moderate with winds North, and with the aid of our powerful slip

light, I at once housed the lifeboat." The fishing boat had a crew of two. The lifeboat towed her into Newlyn Harbour, and returned to her station at 12.30am. Rewards, £11 9s.

In those days, the launch of a lifeboat was a popular spectator event, with large crowds gathering around, and even inside, the boathouse. *The Cornishman* commented on this launch: 'The lifeboat was called to the rescue of the Newlyn steam trawler *Pioneer* that had broken her main shaft and lost her propeller when between Lamorna and Penzer Point. Although there was a quick response to the sound of the rockets by the crew and their reserves, there appeared to the large crowd of onlookers to be some confusion, which seemed to delay the launching and which brought angry comments from the spectators. Fortunately in this instance there was no loss of life, but there are occasions when a delay of even ten minutes might make the difference between life and death. Mr. Hendy, the skipper and owner of the *Pioneer* and his mate, Mr Edwards, praised the prompt action of the coastguards and lifeboat crew. Mr. Hendy added he had been fishing for 40 years, and that was the first occasion that he had been towed to that port.'

In the 1920s, *Pioneer* was used to run excursions from Penzance to Lamorna Cove. She went to Falmouth to muster for the Dunkirk evacuation, but was not accepted, although she completed some war service doing various jobs for the military. Her steam engine was replaced in 1947 by a 44hp Kelvin motor. This was replaced once more by a 30hp Lister in 1952. *Pioneer* continued fishing out of Newlyn until 1966 and was later used as a diving boat for crayfish and salvage work on wrecks. In 1985, *Pioneer* was laid up in the Old Harbour at Newlyn where she stayed, afloat, for seven years. In November 1991, after restoration under a new owner she was relaunched in September, 1999 into Hayle Harbour. The boat is listed on the National Historic Ships Register

1939 JANUARY Early in the morning of **January 21, 1939**, the Belgian trawler *Paul Therese*, of Ostend, broke from her moorings in Newlyn Harbour and drifted out to sea. Her crew of six were asleep and unaware of their danger. A South-Westerly gale was blowing, with a rough sea, and the weather was thick. The news was received from the Coastguard, and the *W&S* was launched at 4.35am. She found the trawler near the rocks between Penzance and St. Michael's Mount and one of the lifeboat crew boarded her. He roused her crew and the lifeboat towed her out of danger. Her crew then got the engine going and she followed the lifeboat clear. She returned to Newlyn Harbour, and the lifeboat arrived back at her station at 7.15am. 'Property Salvage Case'.

A few days later, the *WMN* reported: 'The Belgian trawler *Paul Therese* has been arrested at Newlyn and is being held under the Admiralty Marshal. A few nights ago the *Paul Therese* was, as reported in our columns, rescued by the Penlee lifeboat after she had drifted out of Newlyn Harbour into Mount's Bay. Unaware of their danger, the crew slept. Payment is now being claimed for the lifeboat's services. It is emphasised that the arrest of the trawler is more a matter of formality than anything else. While her crew slept soundly, tired out by a long battle with the sea, the Belgian trawler *PT*, of Ostend, broke her moorings and went stern first from Newlyn harbour on Saturday morning. Drifting to sea she was blown across Mount's Bay and was on the point of striking a reef when the Mousehole lifeboat managed to take her in tow. A member of the crew of the *Hubbastone* [a Bideford owned steamship] said: "The trawler was alongside us. I saw her jerk away from the quay and part her moorings. I shouted but the crew did not hear me."

The Belgian trawlers Paul Therese and Jeannine alongside the quay at Newlyn. Both boats had needed the services of the W&S

Capt J.T. Richards, Harbour Master, St Ives, told the *WMN:* "It happened fortunately that the dock staff were on duty at the time. With great enterprise some of the lifeboat shore staff followed the boat round the shores of the bay in a motor-car and they reported the matter to me at Penzance. I boarded the Trinity House steamer *Satellite*, knowing one of the crew, Mr F.T. George, was an ex-Naval signaller. He came up on the harbour wall with his Morse lamp. We saw the lifeboat searching round the bay, and as she swept her searchlight to and fro we caught sight of the trawler between her and the shore. The trawler had come across the bay between the Gear Rock and Penzance bathing pool. Mr George signalled the trawler's position to the lifeboat and she proceeded to the spot. It was a very fortunate thing this communication was established otherwise the situation might have been more serious."

Talking to the *Daily Mirror*, Coxswain Frank Blewett said: "It was pitch dark and blowing a gale when we launched the lifeboat and went in search of the trawler. After nearly an hour searchlights picked out the vessel, which was about to be driven onto the dreaded Raymonds Reef [. . .] Imagine their surprise when I manoeuvred the lifeboat alongside and got a tow rope aboard, towed the trawler out of danger, put a crewman aboard and then proceeded to rouse the crew, who were still fast sleep in their bunks, and knew nothing of what had happened. Had we been two minutes later in finding the trawler she would have been driven on the rocks and the crew drowned as they slept." He took the *Paul Therese* in tow until her engines were started, and she returned to Newlyn under her own power.

'All Ships. All Ships'

In 1937 the RNLI introduced two-way, 'ship-to-shore' radio to its lifeboats. An article in the *Life-Boat* journal by Engineer-Captain A. G. Bremmer, O.B.E, R.N., Superintendent-Engineer to the Institution set the scene:

'Wireless has now been in use in lifeboats of the Institution for eleven years, but the severe limits of its use, and the great difficulties of its use, in lifeboats are hardly understood by those who have seen the spectacular results of wireless in bringing help to vessels in distress on the high seas, and who think that, as a matter of course, all the Institution's lifeboats should be equipped with it.

The use of wireless in lifeboats is solely for the purpose of keeping them in touch with the shore when they are out at sea and are too far away for visual signals to be seen. They have not the duty of picking up messages from vessels in distress. That duty belongs to the shore stations, of which the General Post Office now has thirteen round the coasts of the British Isles. There are also a number of lighthouses and light-vessels, offices of Harbour Authorities, and a few coastguard stations, which have radio-telephony sets both for transmitting and receiving. It is the shore stations which receive signals of distress. These messages are passed through the coastguard to the lifeboat stations, which act upon them. The Institution began its experiments with wireless by installing a wireless-telegraphy receiving and transmitting set in the Rosslare Harbour, cabin motor lifeboat *K.E.C.F.* (ON700), in 1927. This is the only lifeboat which has been equipped with wireless telegraphy. It has not been used in other lifeboats, because of the necessity of carrying a fully certificated operator.

In 1929 the motor lifeboats at Dover, Stornoway and St.Peter Port, Guernsey, were equipped with radio telephony, and a little later the motor lifeboats at New Brighton and Barra Island were also equipped with it. These five lifeboats had both receiving and transmitting sets with a range of 50 miles. At that time the Post Office shore stations had wireless telegraphy only. Then, in 1931, the Post Office equipped its shore stations with radio telephony of low power.

The six lifeboats which had by this time been fitted were the only lifeboats which fulfilled the necessary conditions. They were within fifty miles of a shore signal station, they were lifeboats with cabins, and they lay afloat. They had to lie afloat, so that the mast and aerial could be kept up for regular testing. They had to have cabins to protect the delicate apparatus from the sea. As other motor lifeboats were built which fulfilled the necessary conditions they were equipped with radiotelephony. Up to the end of 1936 nine more had been so equipped. Like the first five, these nine were all cabin lifeboats lying afloat, with the exception of Cromer—a cabin boat kept in a house. In 1936 experiments were carried out with a radio-telephony set in the Cromer boat, to see if, with improved apparatus, it could be used in a cabin lifeboat which did not lie afloat. The results of this experiment have been sufficiently good to justify the Institution in deciding to fit with radio telephony all cabin lifeboats which are kept in boat-houses, provided, that is, that they are within fifty miles of shore wireless stations. This has added another thirteen to the number of lifeboats carrying both transmitting and receiving sets. These have all been fitted during 1937.

In 1937 a number of cabin lifeboats have been fitted with water-tight loud-speakers on deck, so that the cabin has only to be used in sending messages, and these loud-speakers will eventually be used with all sets in the cabin boats. The present position is as follows: Of 53 cabin lifeboats in the Institution's fleet, one has wireless telegraphy for receiving and transmitting, 26 have radio telephony for receiving and transmitting, and another ten will be fitted with it during 1938.'

One handicap which delayed the fitting of radio equipment to lifeboats was that marine radio operators had to have passed stringent examinations, including proficiency in Morse code and basic electrical skills, in order to acquire the relevant licence. The crew member Clarence "Clarry' Williams was originally entered on the crew list as 'radio operator'.

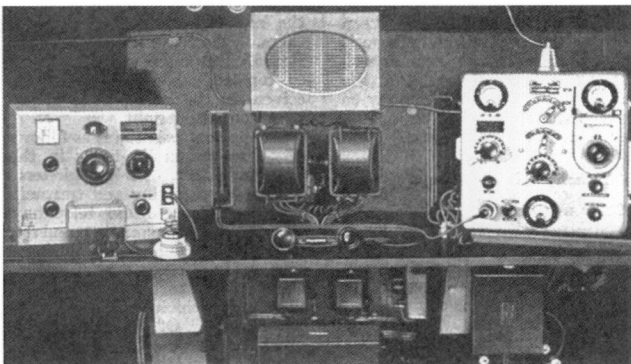

Marconi transmitting and receiving sets as fitted to Cabin-class lifeboats in the late 1930s

Red flares sighted: 'A terrible night'

Two days after the *Paul Therese* incident, at 3.30am on the morning of **January 23**, Coxswain Frank Blewett received a telephone call stating that the St Ives Lifeboat had launched on service to a ship in distress, possibly the SS *Wilston*, NNE of the Longships Lighthouse. At 4.25am Cox'n Blewett received a second message: 'St Ives Lifeboat out on service, signals seen off Clodgy Head. Padstow boat out on service. Sennen boat cannot launch.' He immediately decided to call out his crew, knowing that the Sennen boat could not go afloat due to the weather conditions [apparently, there were rocks washed up onto the launch slipway], and fearing that this was a big job for the St Ives lifeboat. At about 5.00am, in a whole gale and very rough sea conditions, the *W&S* lifeboat and her volunteer crew left Newlyn Harbour and proceeded towards the position given. On reaching a position about 1 mile from the Longships the lifeboat received a recall from Land's End Coastguard. She returned to Newlyn arriving at 8.45am. Weather: NNW whole gale, sea state very rough, showery rain.

How did the boat behave? Splendid. Crew: Coxswain Frank Blewett, Mechanic Johnny Drew, Bowman Luther Oliver, Edward Downing, Clarence Williams, Bertie Jenkin, Harry Blewett, Arnold Gartrell and Ben Jeffery — the Coxswain took one extra crew member due to the horrendous weather conditions.

The St Ives boat *John and Sarah Eliza Stych* (which was on loan from Padstow following the loss of the *Caroline Parsons* one year previously) had launched at 3am into winds gusting to 100mph (160kmh). Along with Coxswain Thomas Cocking were: John Cocking (his son), Matthew Barber, William Barber and John Thomas who had all been aboard the *Caroline Parsons*, along with Edgar Bassett, Richard Stevens, and William Freeman. Rounding The Island the boat met the full force of the NNW storm and capsized but righted itself, leaving five of the crew overboard, while Freeman made it back into the boat. With a fouled propeller the boat capsized and righted a second time, leaving only three survivors. The boat then drifted across St Ives Bay towards Godrevy Point, where it capsized for a third time. When it righted only Freeman was left. He scrambled ashore while the boat was smashed on the rocks at Gwithian. All eight crew members were awarded bronze medals. Since then two more Tommy Cockings, the drowned coxswain's son and grandson, have served as Coxswain on the St Ives Lifeboat.

"It was my worst experience as a youngster, and my first year as Mechanic

Johnny Drew, recently promoted to full-time Motor Mechanic, would never forget that night:

"At 2 o'clock in the morning the telephone rang. 'We're going.' Message from the Coastguard at St Ives. 'Penlee launch. St Ives lifeboat showing red flares.' Now that's danger, when she capsized. She being a small boat and a self-righter, the sea caught her. They are a very light draught, so they haven't got the hold in the water that the original big lifeboats had. Low weighted keel and they are aren't supposed to capsize, they should right themselves; self righting. Well, that was the message, 'Red flares sighted two miles from Clodgy Head'. We got called out and we went. When we were going out you couldn't go on the deck

of the lifeboat. Weather was some bad coming up to Land's End to go into the Channel. Normally the procedure going from Mount's Bay on the west turn, the first sight you would see was the Runnel Stone Buoy all lit up. Now the crew was looking for that, but there was no Runnel Stone Buoy. Couldn't see the light of any buoy. That was gone. So we are going out into the open sea. They knew all the shallow patches around the coast, and one up there which is a very nasty patch called Carn Brea, and that, in rough weather will break up a ship, never mind a lifeboat. So when we got out and we took a bearing from the Wolf Rock we could change now to what we call going to channel, going to the North Channel. Then at last, between the heavy showers, we sighted the Longships light which was white. Now the Longships is a red and white light. When it is red you are on a certain course for Mount's Bay. But you've got to bring it out white, show white to know that you are clear of everything to go in the North Channel. So when we spotted the Longships it was white and we were going into the Channel on a true course.

"Now as we were going, everybody huddled, and I think it was my worst experience as a youngster, and my first year as Mechanic. It was a terrible night. Anyhow, we are going in the Channel. Darkness. Seas terrific. Everything depended on the radio. Wouldn't miss a message for the world. Land's End Radio was about five miles away going up towards the Pendeen light. And at last Land's End Radio was calling us. They had to have the number of the crew. Open the hatch and shut it behind me, to shove me down in the cabin where the wireless was. No remote control, in them days it was operated from the cabin. It wasn't safe for me to go along the deck alone, so someone came with me in the darkness. Got down in the cabin and just as I was about to take the microphone to answer Land's End, the boat rose almost perpendicular and went down with such a severe bang that everything went dead. The radio went dead. Of course, nothing could be done then. It was completely blank on the

LIFE-BOAT SIGNALS

The Life-boat Coxswain is to ensure that while afloat there is always one man specially detailed to look out for signals

SIGNALS TO LIFE-BOAT CREW

NUMBER	MEANING	MADE BY	NIGHT		DAY	
			SIGNAL	REPLY	SIGNAL	REPLY
1	LIFE-BOAT CREW ASSEMBLE.	Life-boat Authority or Coastguard.	2 MAROONS (GREEN STAR) IN SUCCESSION		RED FLAG AND 2 MAROONS (GREEN STAR) IN SUCCESSION	

If for any reason maroons are not used for assembly, a distress signal should be acknowledged by a white smoke signal (by day), and at night by a white star rocket (or white flare if a rocket is not available). The Life-boat Authority is responsible for this only if there is no Coastguard Station in the neighbourhood.

| 2 | I WISH TO COMMUNICATE. Close if necessary and practicable. | Coastguard, Life-boat Authority, Lighthouse or Lightvessel. Latter uses white flare only. | WHITE FLARE OR MORSE PROCEDURE OR 3 WHITE STAR ROCKETS AT INTERVALS OF 20 SECS BETWEEN EACH ROCKET | WHITE FLARE OR MORSE PROCEDURE | Night Signal or Semaphore procedure. | Alteration of course or Semaphore procedure. |

If necessary, maroons may be fired by Coastguard or Life-boat authority to attract attention to No. 2 or No. 3 Signal.

| 3 | RECALL. Return to your Station. or SERVICES OF LIFE-BOAT NO LONGER REQUIRED. | Coastguard, Life-boat Authority, Lighthouse or Lightvessel. | GREEN TURNING WHITE FLARE OR LETTER B BY MORSE — ● ● ● | SIGNAL N° 5 OR IN THICK WEATHER LETTER B BY MORSE — ● ● ● | Night Signal, or Triangular or pendant shaped flag repeatedly dipped and hoisted, or waved from side to side above the head or Letter B by Morse. | Alteration of course In thick weather Letter B by Morse |

WARNING.—A red burgee or red light at the masthead of a vessel, and in thick weather the sound signal MQF ▬ ▬ ▬ ▬ ● ▬ ● ● ▬ indicates that it is unsafe to fire a rocket or tracer over her owing to leakage of inflammable liquid. If in doubt the signal MQH ● ● ▬ asks whether it is safe to fire a rocket.

SEMAPHORE PROCEDURE.—Hoist " J " Flag (blue with white stripe) or wave semaphore hand flags. Repeat each word until acknowledged by " C." End message by " AR."

MORSE PROCEDURE.—Call ships with "A.A.'s, shore stations with " Z "s. Answer with succession of " T "s. Repeat each word until acknowledged by " T." End message by " AR." Acknowledge whole message by " R."

SOUND SPELLING PROCEDURE.—Call with "AA"s or " Z "s. Answer with succession of " T "s. Each word is not acknowledged. If receiver makes repeat sign " UD " go back a few words and continue. End message by " AR." Acknowledge whole message by " R."

NOTE.—A bar over two letters means that they are to be made without the pause customary between letters.

SIGNALS FROM LIFE-BOAT

NUMBER	MEANING	NIGHT	DAY
4	APPROACHING CASUALTY. SEARCHING. REPLY TO SIGNAL NO. 2.	WHITE FLARE	
5	LEAVING CASUALTY. APPROACHING LIFE-BOAT STATION. REPLY TO SIGNAL NO. 3.	GREEN FLARE IF NONE OR PART ONLY OF SHIPWRECKED CREW ARE ON BOARD THE LIFE-BOAT GREEN TURNING WHITE FLARE IF ALL THE SURVIVORS ARE ON BOARD THE LIFE-BOAT	Night Signal or semaphore.
6	MORE AID REQUIRED.	RED FLARE	RED FLAG

This signal will be treated by all concerned as a distress signal and action taken accordingly.

SIGNALS FROM AIRCRAFT

To indicate direction of a casualty an aircraft will cross ahead at low altitude, opening and closing throttle or changing propeller pitch and then head in the direction of the casualty.

A succession of green pyrotechnical lights or signalled green flashes from an aircraft indicates that she has an urgent message to send about the safety of a ship, aircraft or some person on board or within sight.

EMERGENCY SIGNALS (INTERNATIONAL)

F	● ● ▬ ●	I am disabled. Communicate with me.	V	● ● ● ▬	I require assistance.
L	● ▬ ● ●	You should stop. I have something important to communicate.			
U	● ● ▬	You are standing into danger.	W	● ▬ ▬	I require medical assistance.
			Z ▬ ▬ ● ●	Calls shore stations.	

42 GROSVENOR GARDENS, S.W.1.   (Revised October 1954 and reprinted 1957)

By order of the Committee of Management,
A. D. BURNETT BROWN, Secretary

radio. We had a speaking tube in the cabin and I spoke aft to the Cox'n and said what had happened. He said 'You be careful'. So after a minute I pulled myself together and made an attempt to get aft to the crew in the cockpit. Got there eventually and we were, well, up to our waists in water all the time. Some men were looking over the stern, some looking ahead. At last we seen a green flare go up. That was fired by the Tol Pedn Coastguards. The green, everybody watched, and the green turned to white. You had to be aware of the signals. Green to white meant a recall, to come back. Had we not seen that we wouldn't have known what had happened. We came to find out after we got back, because Newlyn Pier was black with people, knowing that St Ives lifeboat was lost with all crew. That was a bad night, with lifeboats out all around the Cornish coast that night for different casualties. **"**

The SS *Wilston* was a UK-flagged, steam powered cargo vessel of 3218grt, loa 102m x beam 14.6, built in 1916 by Robert Duncan of Glasgow, and operated by Miller W.S. & Co of Glasgow. She was on passage from Newport to Tunisia with a cargo of coal and a crew of 30. The *Wilston* was last positively sighted 20 miles NNW of the Longships, however no distress signal of any kind was ever received from her.

A post on Facebook in 2019, by Sheila Finnimore, quoting Old Barry in Pictures, paid tribute to several crew members of the *Wilston* from Barry: Joseph Charles Clemo, John Nilsson (Neilson), George Coulthard and another identified only as Roy ?.

At Tregurthen a farmer found a broken lifebuoy on the beach, one half bearing SS *Wilston* and the other 'Glasgow'. A small piece of timber from the ship was featured in the BBC 's History of the World–Objects.

Just over one year previously, on January 31, 1938 the St Ives lifeboat *Caroline Parsons* had rescued 23 people from the the SS *Alba*,

but as the lifeboat turned for home it was capsized by a large wave. It righted but ran on to rocks. The Coxswain Thomas Cocking and his eight crewmen got ashore safely but five of the rescued men were lost. Cocking was awarded a silver medal by the RNLI and the rest of the crew received bronze medals, but a year later six of them drowned in the wreck of the *John and Sarah Eliza Stych.*

1939 MAY The police reported distress flares at 1.45am on **May 25, 1939,** between Penlee and Newlyn. A fresh NNE wind was blowing and the sea was choppy. Two lifeboat crewmen put out in a motor boat and found the pleasure boat *Jubilee,* of Penzance, with two men and three women on board, only 20 yards from the rocks with her engine broken down. After manoeuvring they took the *Jubilee* in tow and brought her safely into Penzance Harbour. Rewards, £1 10s., in addition to a local gift of 9s.

1939 JULY At 10.50pm on **July 6**, a message was received from CG Penzance of a vessel showing flares apparently in distress off Gear Rock. The *W&S* was launched and after some time found the French ketch *Ster Vras* with her anchor down. The Cox'n went as near as possible and asked if he wanted assistance. He replied that his anchor was holding and he did not require assistance. After finding he was holding well, the lifeboat returned to Newlyn about 12.50am. The sea was too rough to house the boat.

1939 AUGUST On **August 14,** the Hon. Sec. was telephoned by his counterpart at the Lizard lifeboat that Land's End radio station had informed him that a French trawler *St Jean de Luz* was in difficulties off the Lizard and taking water. Land's End Radio confirmed the message which they had received from another station. The Coastguard were contacted and in view of the fact that the Lizard lifeboat was off service, the Penlee boat was launched. At 10.45pm maroons were fired, and in the absence of the Coxswain (away fishing), the Hon. Sec. Barrie Bennetts and second Cox'n Edwin Madron took charge of the boat. At 10.55pm the *W&S* launched and it arrived off the Lizard at 12.15am on the 15th, but found nothing wanting assistance, and altered course from SE to SW. At 12.25 Land's End Radio called the boat to return to station. It was secured to the slip at 2.45am .

1939 SEPTEMBER War was declared against Germany on **September 1**, 1939.

At 9.20am on **February 7, 1940**, the Coastguard at Porthleven reported a vessel apparently in distress, half a mile to the SSW. At the time a WSW wind was blowing, with a rough sea. The *W&S* was launched at 9.40am, and found the Belgian motor trawler *Jeannine*, of Ostend dragging her anchors. Another trawler had already got a line aboard the *Jeannine* and towed her to Newlyn Harbour, with the lifeboat in attendance. The lifeboat returned to station at 2.15pm. Rewards, £8 9s.

The Coastguard reported at 10.50pm on **February 21** that a vessel was ashore near the Penzer Rocks and making SOS on her foghorn. The weather was very dense fog, with a Southerly wind and a moderate sea. Cox'n Blewett summoned the crew and at 11.15pm the *W&S* was launched, and found the SS *Westown*, of London, had got clear of the rocks and dropped anchor. 'I asked the captain if he still wanted assistance and he said he would like for me to stand by him as he was going to try to get his ship into Newlyn harbour. This I decided to do. He at once hove up his anchor and followed me to Newlyn, arriving there about 2am on the 22nd. The sea being too rough to to put the boat on the slipway, there she lay until 3pm February 24th.' Rewards, £20 12s. (Second Cox'n and Asst Mechanic away fishing).

Westown (710grt) was built in 1921 by C. Rennoldson as Channel Queen. In 1939 sold to Brook Shipping and renamed Westown. In 1947 acquired by Holderness S.S. and renamed Holdernook. In 1956 renamed Logholder. Broken up in 1956.

Five days earlier the *Westown* had been in difficulties in Gerrans Bay and had been escorted to safety by the Falmouth lifeboat.

On **March 17** the Relief lifeboat *B.A.S.P.* was on station. As Johnny Drew was not on the crew list for this 'double' shout, he must have been with the *W&S* which was being serviced or overhauled. John Foster was listed as Mechanic.

The relief boat (ON687) was called out with a message that a French ship the SS *Louise Marguerite* 'wants a lifeboat immediately'. The boat proceeded to the position given but in dense thick fog, nothing was found. The boat returned to station to refuel and the Cox'n went out again about 2am and searched until 7.30am but found nothing. On returning to the harbour the lifeboat crew 'were the means of saving another ship the *Miervaldis* of Riga'. Cox'n Frank Blewett sighted her going direct for the rocks. 'I hailed her and directed the captain to go full speed astern. This he did and saved his ship from going on the rocks.' The Hon. Sec. remarked that the *Louise Marguerite* had reached a western port. 'The Cox'n was justified in making the second trip which was apparently of great value to the second boat.'

B.A.S.P. (ON687) built in 1924, was originally on service at Yarmouth, Isle of Wight and then served at Falmouth before joining the Reserve fleet. Since retiring from the service she has been restored and can be seen on display at Chatham Historic Dockyard

W&S stars on the silver screen

During World War ll, which was declared on September 1, 1939, civilian photography and personal correspondence were strictly controlled, as was media coverage of the hostilities. Cinema was a popular form of escape for the beleaguered citizens and a series of documentaries and docu-dramas was produced or sponsored by the Ministry of Information for theatrical projection, supporting the main feature. One of these morale-boosting propaganda films, titled *SOS: A Cornish lifeboat answers a distress call*, shows the *W&S* and her crew in action. An opening statement reads: 'This is the story of a lifeboat and the men that man her, fishermen of Cornwall. But it is a story that could equally be true of any other lifeboat crew round the shores of Britain.'

Shot in black and white, *SOS* was made in 1940 by Tida Films and

Scenes from SOS: the Coastguard at Treen; the crew going the wrong way to work; the lifeboat afloat with crew stepping the mast; Jack Wallis, Ted Downing, Luther Oliver with Cox'n Blewett facing them; rescue of unnamed crew and footage of a shipwreck BRITISH COUNCIL

is now available from the British Council. Directed by John Eldridge, this 12-minute silent film had a voice-over commentary. The dramatised documentary was structured around a narrative in which the *W&S* is visited by an RNLI inspector who, on return from exercise, sees that the starboard propeller is damaged, and a new one has to be ordered from the RNLI stores at Elstree.

The film opens with a lone fishing boat (PZ 302) heading home to Mousehole, whose crew report they have lost contact with another boat. The Coxswain, Frank Blewett, has to decide whether to launch the *W&S* to search for the overdue vessel. The Coastguard gives the order to man the lifeboat and stand by, illustrated with atmospheric shots within the boathouse as the crew prepares to launch. However, the missing boat has now been seen and the crew are stood down, while the commentary emphasises that 'Even as Mousehole sleeps, lifeboat men will be on the alert as they know that out there at sea other ships and other seamen might be at danger. Even over this peaceful scene looms The War.'

The next sequence shows the crew preparing for a routine Inspection. The *W&S* is described as 'virtually unsinkable, she is as efficient in her lifesaving work as scientific ingenuity will allow'. The boat is launched down the slipway with the Inspector on board. On return to the station, Frank Blewett and the Inspector notice the starboard propeller has lost half a blade which, the commentary insists, must be replaced *'AT ONCE'*. A replacement prop is ordered from the RNLI Headquarters at Elstree and the spare part's journey is followed by van to Paddington station and thence to Penzance and on to Mousehole, just as a British tramp steamer sends out a Morse Code SOS call for assistance and is seen drifting in heavy seas.

By a stroke of luck (and cinematic contrivance) the replacement propeller is being fitted while the lifeboat crew are seen running through the streets and lanes, banging on the doors of their colleagues. More cinematic licence shows the crew running the wrong way through the village, away from the boathouse — presumably for composition or lighting requirements. The station Mechanic quickly fits the new prop, and the boat is made ready just as the crew arrive. But in the next shot our sharp-eyed researcher has spotted that the boat shown launching from Penlee is not the *W&S*, but actually the *B.A.S.P.* which would have been relieving the *W&S*, presumably out of service while the prop is repaired. This gives a clue to the possible date of the filming.

A fictionalised sequence then shows library footage of a steamer

under heavy weather with the ship's crew being transferred to the lifeboat. Radio comms are given prominence and a final shot shows the deck of a sinking cargo vessel with the Wolf Rock lighthouse in the background. The voice-over commentator concludes: "Not for the first time Frank Blewett and his crew have risked their lives to save the lives of others. And with the weather clear again these lifeboat men return to their normal jobs of fishing."

The *W&S* was launched on **July 24,** following a phone call at 3.35am, **1940 JULY** to search for an unknown bomber that had crashed into the sea 2 miles SW of Treen lookout. After making arrangements with a Naval officer, the lifeboat was launched and searched over a large area but nothing was found. Returned to station, arriving back at 7.45am.

Exactly one year after the start of hostilities, the war arrived in Penzance with a shock on **September 1, 1940**, when the minesweeper, HMT *Royalo*, a former Grimsby trawler (GY941), was blown up 1 mile SSE of Penzance harbour. Cox'n Blewett recorded: 'At 12.23pm I received a telephone message stating that a minesweeper had been blown up off Penzance harbour. I at once got the crew together and prepared for launching. Then I got a second message and I launched the lifeboat at once and proceeded to the position given. When I arrived on the scene I found that small boats had picked up survivors. I stayed around the area of the casualty for a while in case any bodies should come to the surface but found nothing. I returned to station, arriving back around 2.20pm.'

Writing on Nostalgic Penzance & Newlyn website, George Barnes recalled: "I was a young boy, nine years old at that time. I saw this event happen and it has always been in my memories. I was on Larrigan rocks, playing with my friend when there was a huge explosion, and looking

HMS Dee was an armed trawler similar to the Royalo. These ships were crewed by Royal Navy men and carried anti-aircraft guns

towards the Gear pole we saw this huge plume of black smoke with masses of debris flying up in the air. As it subsided there was a second explosion, and after a few minutes there were many small craft heading towards the site of the explosion. The crew never had a chance [. . .] Apparently she hit the mine which lifted the bow out of the water, as she settled back she hit a second mine, moments later she disappeared below the water.

The *Royalo* was a steam powered trawler built by Cook, Welton and Gemmell in Beverley (Hull) in 1916 (yard no: 334), of 248 grt, 35.7 m x 6.7 x 3.7 draft, powered by an 80 horse power, 3-cylinder triple expansion engine, single shaft and propeller, capable of 9.5 knots. Delivered to George F Sleight of Grimsby with number GY941, who owned the vessel through until 1939, although she was requisitioned by the Admiralty from 1916 to 1919 as HMT *Royalo* (FY2995).

Requisitioned again in November 1939 at the start of Word War 2 as an auxiliary patrol vessel (APV) based in Lowestoft with the number FY825, she was subsequently used as a minesweeper and was assigned to duties in Cornwall. Mount's Bay had been mined by German aircraft on the previous day, and the steel hulled vessel struck a magnetic mine one mile off the Penzance waterfront. Most of the crew were trawler men serving with the RN Patrol Service, although some were Royal Navy Reservists, including the Skipper William Warford.

The *Royalo's* initial crew complement was 12, but the Penlee Cox'n report states there were 19 persons on board, eleven of whom were lost: Robert W.E.G. Burgoyne, (64) Engineman of Northumberland; Henry Thomas Dukes, Engineman (45) of Grimsby; William Henry Greenfield, Stoker; Sam Lockwood-Dukes (22) Stoker of Worsborough Bridge; Raymond Ormerod (20), Telegraphist of Wroxham; Jim Walker Pitts (28) Engineman of Great Yarmouth; Leonard Rye (26) Second Hand of Hull; Thomas Gardner Taylor DSM (21) Signalman of Newcastle-on-Tyne; Robert John Tilley, Seaman (27) of Whitstable; William Durrant Warford

Results of the CISMAS survey on the *Royalo*. Built as a steam trawler, *Royalo* was commissioned by the Admiralty during WWI and in WW2 as an auxilliary patrol vessel. She was working as a minesweeper when she hit a magnetic mine and sank near the entrance to Penzance harbour in 1940. The Royal Navy cleared the wreck using explosives in the 1960s. A surprising amount of wreckage lies on the seabed but much of it is buried under sand.

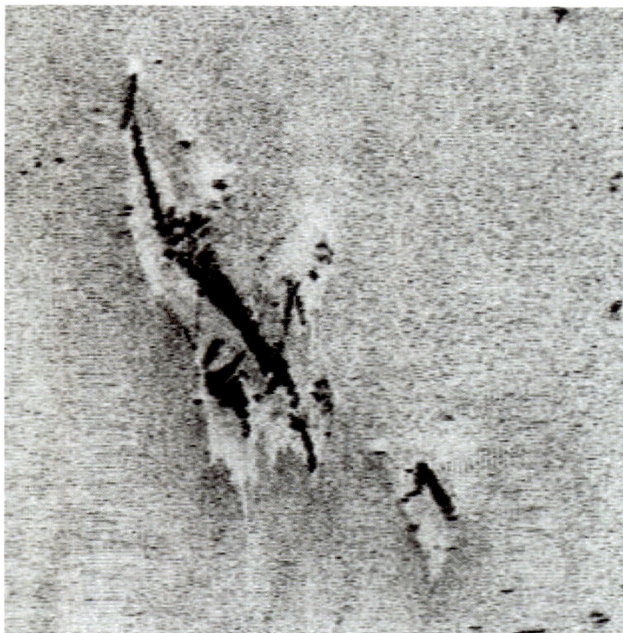

Target No	A807
Position	320072E 5554272N
Chart depth	8m
Vessel name	Royalo
Vessel type	Steam trawler
Built	1916
Length	36m
Tonnage	248
Wrecked	1st Sept 1940

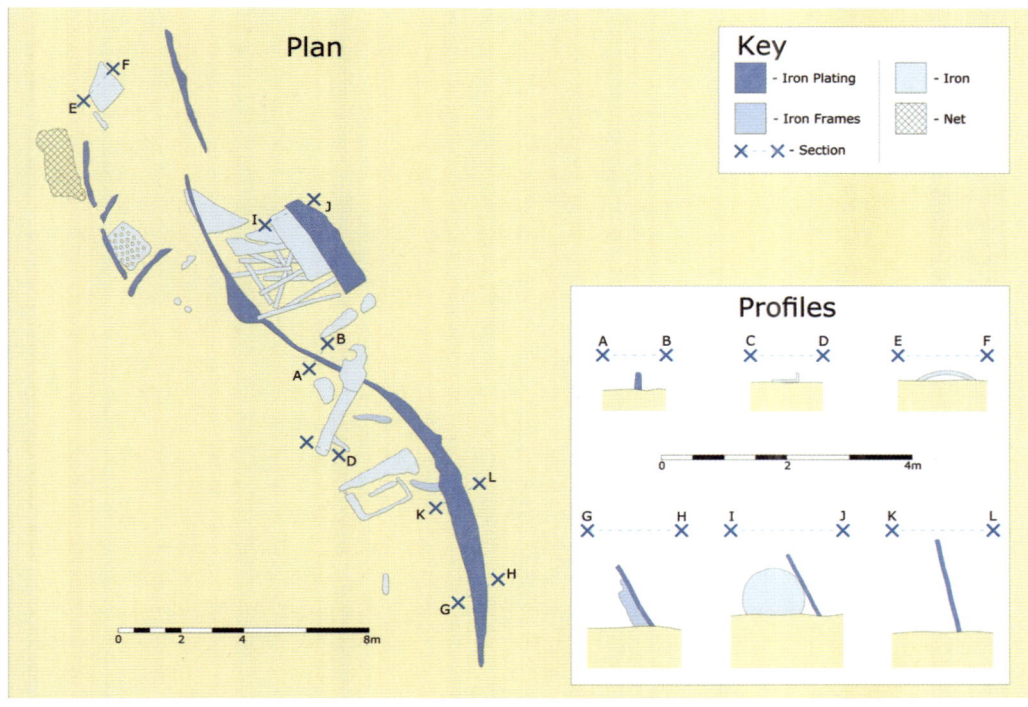

Plan

Key

- Iron Plating   - Iron

- Iron Frames   - Net

X - X - Section

Profiles

DSC (45) Skipper of Pakefield; Irvine Willox Watt, Lieutentant RNR (33) of Kensington, London. Six of them are buried at Penzance cemetary

In addition to more than 2,000 class vessels built for the Navy there were 215 requisitioned trawlers in WW2 of no specific class, 72 of them were lost. Some 1,250 trawlers of all types were sunk and 15,000 personnel killed. These ships were armed with Anti Aircraft and machine guns. As recently as 2007 divers in Mount's Bay have advised to be aware of ammunition around the *Royalo* wrecksite.

The wreck lay there for many years marked by a green wreck buoy In the mid-1960s the Royal Navy 'dispersed' it with explosives.
The remains of the wreck lie at 50.06.46N, 05.30.56W

The War Years

Throughout the war years the whole emphasis of lifeboat service had changed. Recently employed as full time Motor Mechanic, Johnny Drew had first-hand experience of the lifeboat's new role:

"When the war broke out all the lifeboats came under the ruling of the Admiralty. Newlyn was a Naval base, a Sub-base. A lot of these fast inshore rescue boats were there, called MLs [Motor Launches]. Their speed was terrific, same as air sea rescue, because that's what they were for. And normally they took over a lot of jobs that the lifeboat would have done. The lifeboat was

ML 542, based locally, was in action during the submarine incident at Wolf Rock (page 130). Photographed at Hugh Town, IoS in1943 by George Baker
© GEORGE BAKER

always kept as the last resort. The lifeboat would never have been sent out during the war years unless really essential, if something happened that the other boats couldn't go to. The lifeboats were supervised by the Admiralty but had to carry out instructions such as in peace time. If somebody reported a boat in difficulty: 'Phone the lifeboat authority and the boat will launch'. But during the war years, No! Because all around this coast and in the Bay was all full of magnetic and acoustic mines, all dropped. All the waterways around the British Isles were done, Falmouth and all. The enemy sent their bombers during the night time and dropped these mines in the waterways and bays to stop our shipping, you see. All dropped by parachute [. . .] dropped to blockade our ports. So, the Navy had their spotters out to plot actually where these mines were dropped. Therefore, they knew the area.

I remember one evening, it was a perfect day, fine beautiful weather. I was out at the lifeboat house and you could hear when the air raid warning would go. But on this occasion no air raid warning went off. Now, there were troops all around Newlyn, a vast number of troops all around the bathing pool. That was a fortress, they had big guns in there and searchlights, just in case of invasion. They were manned 24 hours around. Anyhow, this particular evening around tea time, everything was quiet. The Germans had occupied the Channel Islands by then. They had their base there. Well, this single plane had come along and was flying so low that it was below the range of radar. And when it came in it entered Newlyn harbour lower than the lighthouse. When it came along the shore no air raid warning went. And when it approached Newlyn harbour and came through the lighthouse as its mark, it heightened above the lighthouse, flew over the harbour, and there was a big coal hulk there [. . .] it dropped two bombs, dropped one bomb on that, nobody on it, just the coal, then he opened up his machine gun fire and killed one chap, George Basil Chiffers. Up above the Newlyn slip there, chap was walking along. Well it scared everyone. Before you could say Jack Robinson it was gone. We didn't think that was going to reoccur, but it did.

The next night the same thing happened. I was at home then, around the same time. The air raid warnings went, the siren went. This plane had gone in, exactly the same tactics. Of course, the barrage had him, guns everywhere, firing. I lived at the Gurnick, down the other end from the lifeboat house, standing there crouched behind a wall and seeing this plane coming out on fire. They had three Naval patrolling boats out there, standing by outside. One behind the island, so if he escaped they could open fire on him. He was hit and on fire. His machine guns were going and I was there protecting myself by the side of the house, seen it pass over and one of the patrol boats hit it again. It came down just off Penzer Point. Of course, one or two Naval craft were sent out, but I can't tell you the full details because it was in secrecy. But the two German airmen were picked up and taken back to Newlyn.

During the war we had to be on standby all the time. As I said, in those days you were controlled by the Navy, all your orders came from them. With the system they had, the Coastguards, through the Navy, would ring the Coxswain. But the base would ring me. So I was always rung up by the Commander of the base, Commander Walsh. I had a lot of secret papers handed to me and I had to keep secret myself. Know what to do in case of

"During the war we had to be on standby all the time. . . you were controlled by the Navy

invasion, invasion by air. What I would have to do to stop the lifeboat falling into enemy hands. What to do with the machinery so it wouldn't be workable.

On the Dunkirk affair [May-June, 1940] a lot of the inshore fishing boats could go out by day but were not allowed out by night. During the latter part of the Dunkirk operation, they were called out, along the south coast, from Dover to the Southwest. Lifeboats went from Dungeness down to Shoreham and, of course, even the Thames launches went. Boats here were taken over by the Naval authorities. They were refuelled and they had to go to Falmouth. Then they would be towed by Naval craft to the beaches. Well, that wasn't put into practice for Dunkirk because they had sufficient boats. The following week or two there was another scare for Brest. Then the plan was put into action. Every available boat from Newlyn, Mousehole, Porthleven, St Ives, all the boats that were day fishing were all commandeered and proceeded to Falmouth. They waited for orders to be towed across and remained there a day or two. The crews were released then and they came home by rail, leaving the boats there. My fishing boat *Bonnie Lass* was one of them. After it was called off the Naval authorities wouldn't deliver her back, you had to make arrangements to go and get her.

At the start of the war when the enemy invaded Belgium, the whole fleet of Ostend trawlers, which were beautiful boats, they all left Ostend with their families. They came across here to Brixham, Plymouth and Newlyn. Newlyn was full up with all these families. A notice went out for anybody with anything to spare in the furniture line, to give or loan to accommodate these people who had lost everything. There were six or seven families living in Mousehole. During that period, when they first arrived, all fishing was banned, day or night.

No Fishing. These beautiful trawlers were laid up. And there were so many at Newlyn. Mousehole harbour, on account of possible invasion, had been closed. Booms were put down to stop everything. Eventually the Admiralty took over the harbour and patrol duties. All these beautiful trawlers had to berth in a safe harbour for the time being. So, Mousehole had 14 in this harbour, all along the piers: dropped anchor and laid up. They were here for perhaps two years or more until the Admiralty had use for them. They would take them in command and if their owners wanted to go with them they were taken on as well. We had all the Belgian trawler fleet all around the coast, up to the north, and up Scotland, serving our ships with harbour duties and patrol duties. They were lovely boats. They came with their families. **"**

1940 NOVEMBER On **November 25** a Belgian trawler *Marguerite Simoune* was sunk by gunfire from German motor torpedo boats at 12.30am, while fishing in Mount's Bay. Her crew were picked up by another Belgian trawler *Roger Denise*, and landed safely at Newlyn. It was also reported on that day, that a small convoy of two steamers and a tanker, from Plymouth to Avonmouth, was attacked by German destroyers in Mount's Bay. The Netherlands flag vegetable oil/wine tanker *Apollonia* (1931-built, 2086grt,) was sunk by the *Karl Galster*. Fourteen crew were killed (Captain P. Schol) and there were 20 survivors. The UK steamer *Stadion II* was left afloat, when the German destroyers left, but not seen again. After a naval and air battle some 10 miles SE of the Wolf lighthouse, the lifeboats from Sennen Cove, Penlee and Lizard were tasked to search for survivors but nothing was found except a large patch of oil from the *Apollonia*.

The Penlee Cox'n received a message at 5.30am from the Coastguard at Penzer Point. He launched the *W&S* and, after searching as directed, he later spoke to one of two Naval patrol boats that had arrived on the scene, but could not gather any information. The boat returned to station making a zig-zag course, arriving at the slipway about 1.15pm. *Apollonia* went down at *50.03N, 05.30W*

1941 FEBRUARY At 5.17pm on **February 2, 1941**, the Coastguard reported a vessel was in distress, and the Sennen motor lifeboat *The Newbons* was launched at 6pm. A strong NE gale was blowing, with a very rough sea. The Sennen lifeboat found the SS *Heire*, of Oslo, three miles WNW of The Brisons. She had lost her propeller on passage from Dartmouth to Port Talbot in

The *Heire* was delivered in June 1917 as *Stend*. (896 gt/1250dwt,) In 1924 renamed *Heire*. Left Norway in 1940, spent one week off Omaha Beach during Normandy invasion. Renamed *Svanholm* in 1949 Modernized in 1951 and from 1963 as Panamanian *Bamby*

ballast, was waiting for tug, and did not need the lifeboat's help. The lifeboat returned to her station at 8pm and stood by. At 11.40pm a message came that the steamer had fired a red rocket, and was thought to be ashore on the Shark's Fin, near the Longships. The weather had now got worse. The Sennen Cove lifeboat could not be launched, and the

Penlee lifeboat station was informed, through the Coastguard, at about 11.30pm, and the *W&S* left at midnight. She found the *Heire* in a very dangerous position near the Longships lighthouse, about five or six hundred yards from the rocks and labouring heavily. Cox'n Blewett spoke to the captain about abandoning the ship but he said he was still waiting for a tug, and the *W&S* stood by her until 9 o'clock next morning when the tug arrived. The lifeboat continued to stand by until the *Heire* was safely in tow towards Falmouth, and then returned to her station, arriving at noon. She had been out over 12 hours. It was a long and arduous service in bitter weather 'very cold, hail and snow' and the Penlee crew were thanked by the resident naval officer at Penzance. A gift of £16 was made.

On **February 25**, the Cox'n's log shows the DI visited the boathouse.

A total of three services were rendered in one day on **March 8**. The Cox'n Frank Blewett's Returns of Services records that a message was received at 2.28pm from the District Officer Coastguard to proceed to the steamer *Margo* (7160grt) of Cardiff, in ballast, off Mount's Bay. The Cox'n arranged with the captain to come into the Bay and drop anchor. The *W&S* came alongside and brought ashore three wounded men and a body from the *Margo*,

then returned to the ship to put a doctor on board to attend to another severely wounded man.

On the same day one wounded seafarer was landed from the *Falkvik* of Solvesborg. The lifeboat took a doctor to her and afterwards landed a man with an injured hand at 10pm. She then took the captain of the *Margo* back to his ship.

A headstone in Penzance Cemetery marks the grave of John Ostrich, who served as Mess Room Boy on the *Margo*. He was aged only 14 years and 344 days at the time of his death. John was the son of Louis and Nancy Ostrich of Canton in Cardiff.

An account from a Merchant Navy Message Board, by a guest who signed in with the name of SIF9HD8. 'On the afternoon of March 8, 1941, sailing in the English Channel, the *Margo* came under attack from three German aircraft who proceeded to rake the ship with machine gun, cannon fire and bombs. Although no bombs or explosives hit the *Margo*,

the ship was violently shaken by the concussion of the near misses and her hull and superstructure were pierced by cannon and machine gun fire. Crew members returned fire with small calibre weapons, and in the process hit one of the aircraft, which was subsequently seen to break off the attack and black smoke was observed coming from the starboard engine. The remaining aircraft continued their attacks for several more minutes, which was eventually broken off and the aircraft disappeared over the horizon. While assessing the ship's damage, it was found four crew had suffered various injuries and the young Mess Room Boy lay dead. A course was set for Penzance to land the wounded and the dead.' The weather was too hard to rehouse and the lifeboat lay at Newlyn until March 10, when it was brought back to the boathouse. 'Second cox'n away on Naval duties.'

There were no more shouts for the *W&S* for more than a year, with RN and RAF craft taking on rescue duties, but the lifeboat crew were occupied with frequent exercises and movements of reserve boats.

1941 MARCH A relief boat arrived on passage for Padstow on **March 13 1941**. The boat left Newlyn for Padstow March 14.

On Monday, March 17, the deck log shows a mine exploded off the slipway about ½ mile NE by N.

On **March 25** members of the Penlee crew left for Appledore to fetch a reserve lifeboat, *M.O.Y.E.* (ON675). They left Appledore the next day for Padstow, arriving at 5pm.

On **March 27** at 8.30am they left Padstow, arriving back at Newlyn at 5pm. The reserve boat was kept at Newlyn and visited often. Presumably the *W&S* was being overhauled or repaired at the Penlee boathouse.

A whole gale lasted until midnight on March 31. Crew assembled to stand by at 1.15pm

On **April 24**, the reserve lifeboat ON675 left station at 11pm to search for an aeroplane reported down 5 miles SW of Portleven, but found nothing. Returned to Newlyn harbour at 4.45am.

1941 APRIL The *W&S* was back in service on **April 29**. The deck log reported the 'Penlee lifeboat ready for service, and transferred gear from the Reserve boat'.

Another relief lifeboat arrived at Newlyn on passage on **May 1**, at 3pm. Left on May 2.

On **May 11**, crew arrived to take over Reserve boat ON675. Boat left on May 12 for Salcombe.

Exercise of *W&S* on **May 30**, 'Everything OK'. Exercise for engine trial on **June 20**. On **June 26**, lifted moorings and found buoy worn out. Quarterly exercise on **July 2**, with full crew and District Inspector. 'Everything OK'.

September 24, put out buoy and moorings. **September 30**, exercise with full crew.

Exercise for engine test with DI on **October 6**. 'All OK'. Another exercise on **October 16**. Visit of DI on **November 14**. On **November 17**, a new compass arrived, and was 'fixed' the following day. On **November 26**, visit of hull surveyor. On **December 2**, the boat was launched for engine test and compass adjustment.

1942 APRIL The first service of 1942 was on **April 24,** when the *W&S* once again went to the aid of three vessels. At 6.30am the Coastguard reported a fishing boat in distress SW of Penzer Point, and the *W&S* was launched at 6.50am. An ESE gale was blowing, with a rough sea. The lifeboat found the boat about 300 yards from the shore and took her in tow. As she passed Penzer Point the Coastguard signalled to her that another boat was in difficulties off Treen.

After towing the first boat to safety, the lifeboat returned to help the second boat, but on the way saw a third boat in a very dangerous position near the Runnel Stone Rocks. She went to her at once, found that her engine had broken down, and took her in tow. She then picked up the second boat and towed them both into Newlyn Harbour, where she arrived at 1pm. The three rescues had taken over six hours. The fishing boats, which all belonged to Newlyn, were the *Margaret*, *Boy Don* and *Alsace Lorraine*, and they were manned by French refugees, 12 in number. An increase in the usual money award on the standard scale was granted to each crew member, and to each helper. Standard rewards to crew and helpers, £8 1s. 6d. Additional rewards to crew and helpers, £4 7s.; total rewards, £12 8s. 6d. The Hon. Sec. noted 'Appears to be a meritorious service under strong weather and difficult conditions.'

The boat was launched at 10am on **October 1** for its regular six-weekly exercise, with Commander Drewry present. 'Tested all lights and reverse gears. Tested radio with Land's End and Newlyn Naval Base which proved very satisfactory. Returned to slipway about 11.30 and rehoused boat. Boat and winch engines satisfactory.'

There was another exercise on **November 17**, with Mr Davery (Surveyor of Hulls) present. Mechanic John Drew 'tested the radio with Land's End and Newlyn Naval Base on both wavelengths, but Land's End reports on 181.8 metres were rather low about 178 metres.' He then checked 181.8 metres with Newlyn Naval Base, who reported as correct. After a telephone conversation with Land's End, he informed Marconi, Falmouth. Those Motor Mechanic's Records that survive show that the radio was often giving problems. Drew's attention to the engines showed they performed satisfactorily on all his reports.

BRITISH COUNCIL

On **December 6** the *W&S* launched for engine test and compass adjustment.

The *W&S* was launched on **December 15** to search for an aircraft rubber dinghy reported one mile South of Wolf Rock. There was a whole gale blowing SW to S, and the sea was very rough with rain and bad visibility. Cox'n Blewett 'Arrived on scene at about 11.30pm, searched about, cooperating with aircraft which dropped flares for approx one hour. Aircraft returned to base and lb continued search until 4am and then left vicinity and proceeded towards station.' After approximating position, and conditions being very bad, Cox'n Blewett called Land's End Radio to request Penzer CG and Searchlight Company at Penzance to show their lights. Assisted by the lights they found and entered Newlyn harbour at 6.30am. The dinghy had not been seen. 'Second Cox'n away fishing (Nicholas Richards acting), bowman sick in bed.'

The Hon. Sec. remarked: 'Weather as bad as any the Coxswain had experienced. On return, Newlyn harbour was indiscernible and the Coxswain showed initiative in calling upon Searchlight Company, who responded with alacrity. Cox'n informs me their lights made all the difference and enabled him to make harbour with certainty. I have written to CO of Searchlight Company conveying thanks. The MM (Mechanic) reports that in connection with the radio, the drive valve of the transmitter jumped out of its base. This was due to very rough weather. A similar result occurred on another occasion in similar weather. I wish to draw your attention to the good work of the Mechanic who, under great difficulty, replaced the valve, and subsequent contact with Land's End was good. For very rough weather the receiver aerial is too slack and requires attention. The Motor Mechanic informed Marconi and they have promised to attend to it as soon as possible.' The Mechanic in question, Johnny Drew, painted a grim picture of the night's service.

      "I know the night very well because it was coming up to Christmas and I was at home with my wife and sister, the blackouts up. The telephone rang, about 9 o'clock, it was the chief of the Coastguards. I knew him well. 'Johnny my son, I've got a message. You are going to have a very bad night. Get your crew ready and proceed to the Wolf Rock Lighthouse. You are going to have a rough night, but it's reckoned that one of our bombers has crashed out there somewhere in that area.'

      In those days the Wolf wasn't lit; it only had candle power. When they knew of an approach of a convoy they would light up for a short period. The wind had moderated a little but the sea was up, terrific sea. We were

"Because the wind and the sea was so furious you feared for your own safety

proceeding at full speed out to the Wolf, timing her because we couldn't see no light. When we came to an hour and 50 minutes we knew we were alright with timing and at last we see the glimmer of the Wolf. Well, no lights, blind and could hardly stand on the deck. So we were on station and the only thing to do was perhaps shouting and hope to hear some reply. Had a conflab with Cox'n Blewett, what to do. From the time we were called it was three hours now. With the drift the airman would be up towards the Sevenstones lightship. Well, we were going before the wind and sea then, up towards the Sevenstones, so we were steaming at full speed. You could go out and look then, because you were running with the seas, they weren't coming over you. It seemed ages we out there, for about an hour. We were getting in the area of the Sevenstones, which we knew because we started running to shallow water, and what we call the groundswell was getting more noticeable. Slowed down and kept watching and shouting. I shall never forget, it was getting around midnight and all of a sudden out of the skies came three big chandelier flares. Lit up all the sky like day, for miles around. Aircraft searching.

      We were dodging then, set the course back to the Wolf where we started. At last the flares died out, nothing seen and nothing heard. So as we were dodging back, just gone midnight and the wind freshened so quick, it was almost unbelievable. Because the wind and sea was so furious you feared for your own safety. We turned and tossed and dodging to get to the Wolf. After an hour or so steaming at that speed we got eventually by the Wolf Rock light. Then we had to decide what to do. So we were all under the canopy, because you dare not expose yourself. The weather now was furious, under water like a submarine. Water was breaking everywhere. Well, if we had to stop there until daylight before we started to run into the Bay, I don't know what might happen. We figured it out, the Cox'n and me, and he was in charge. He showed a lot of authority at times. No lights to steer on and coming into a blackout now in the Bay, with a most severe gale that we were having. He said, "As soon as we turn on, Johnny, you go down the cabin and tell the authorities [on the radio] we are now leaving. Leaving our position to return to base. Now, from now." So I came up after sending the message and turned the engines on full. Now the wind was South South East, hitting us from the side. We had to steer East, East to Mid North. The seas were hitting us and we had to go through it in the darkness, under water the whole time, the lifeboat was. Of course, the hatches were all shut. We steamed at full speed and everybody had their lifelines on. The Cox'n said "Hour and 20 minutes and then we are going to heave to. Not going any further. That has given us our rough time, we'll be in the Bay here somewhere."

Cox'n Blewett was in charge. 'He showed a lot of authority at times.'
BRITISH COUNCIL

We steamed in and at last he said "Slow down". It was blowing so hard you couldn't stand on deck but he said to me "Go down the cabin and call them up, and ask them." I went down and called Land's End Radio and you had to be very brief. "Penlee lifeboat now hove to. Will you ask the authorities to turn on the powerful lights." I wasn't out the cabin before the lights were on. When I came up they were searching the Bay and they picked us out. We were about two miles off Penzer Point.

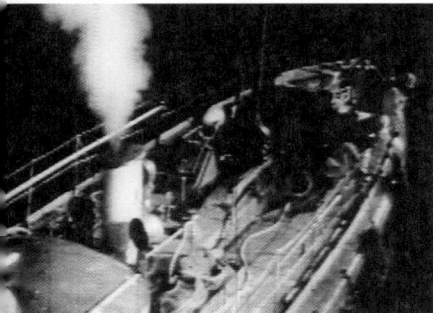

The weather! When I came out and shut the hatch I was standing on her deck, water to my knees. As soon as the lights came on, Cox'n set the course to Newlyn harbour. We turned the power on full and came in on the beam of them lights. I shall never forget, when we got into Newlyn harbour, the Commander in Chief was there and a lot of Navy personnel. We were soaked through, hard to get up the ladder to go and have a cup of tea in the hut there. It was blowing! There was a prominent man around here who was connected with the Home Guard, he said to me that was as bad a time as any lifeboat had. As I was employed, my mind was working so I can't say I was frightened, but might be rather scared. I never gave it a thought that I might die as I was too busy, thinking of the radio or concentrating on our engines. Time seemed to go so quickly and that hour and 40 minutes from the Wolf Rock to the Bay was furious. Because we were taking all the weather on the beam, the side. Oh some sea! The Southerly sea is a shorter sea, a breaking sea over her. It is a shorter sea than a South Westerly sea which is a longer one. A Southerly sea is so rapid, one after the other. We had a buffeting that night!"

The first service of this year was on **June 10, 1943**, when the lifeboat 1943 JUNE was launched at about 1pm, to search for an overdue fishing boat *Sunshine* with a crew of five. The Cox'n and Second Cox'n were out fishing, so Acting Cox'n Luther Oliver and Acting Second Cox'n Bertie Jenkin launched the lifeboat. The vessel was found in tow of a Belgian trawler and lb returned to station. Before launching the Mechanic was informed by RNO (Royal Navy Operations) that all messages by radio were to be made through Land's End.

1943 AUGUST A message was received at 06.30 on **August 20** from the Coastguard at Penzer, reporting a Dutch motor vessel *Actinial* flying a distress signal. The lifeboat was launched and found the small coastal vessel had suffered engine trouble, but the Coastguard informed the Cox'n that the ship had overcome its trouble and was proceeding towards Newlyn as the lb stood by. The sea was too rough to house the lifeboat, which arrived at Newlyn abut 8.30am.

1943 SEPTEMBER The lifeboat was launched on **September 14**, to search for a boat capsized near Tater Dhu. After receiving the first message at 7.37pm, the Cox'n waited for further news from the Coastguard. But, as the telephone line was engaged, he decided to launch the boat, as he had the local knowledge required in that vicinity. He arrived at the scene at 8.20pm, searched about, but found nothing. Speaking to a man on the cliff, the Cox'n was told the last he had seen was four men clinging to a raft or upturned boat about half an hour previously. The search found nothing and, after reporting to the Coastguard by Morse code, the boat was told to return. The sea was too rough to house the lifeboat, which proceeded to Newlyn harbour, arriving about 11.15pm. The fate of those four unfortunates is unknown.

1943 NOVEMBER Unusually, a Royal Navy officer directly ordered the lifeboat to launch at 9.55am on **November 14** to search for a missing landing craft HMLC1. reported South of the Wolf Rock Lighthouse. It was blowing a full gale with 'very rough seas, hail showers and very cold'. Finding nothing at the reported position, the lifeboat tried to contact Land's End Radio, but 'we had something wrong with our set'. The boat searched until 2.20pm and then was instructed to change position. Once again, nothing was found and the boat returned to Newlyn, arriving at 9pm. After 11 hours at sea in terrible conditions a request was made to replace stores: 'One Tin of Biscuits, One Bottle of Rum.' The boat was housed two days later.

When the true horror of war struck home

1944 JANUARY On **January 6, 1944**, the Royal Naval Office at Penzance reported a convoy of merchant ships had been attacked about 6 miles south of Treen Coastguard Hut. At 5.50am, in a moderate SW wind and rough seas, the *W&S* launched from Penlee Point. At the given position they found two liferafts, one with two men, the other with 10 men and two women, all survivors of the Swedish steamer SS *Solstad* of Norrkoping, which had been bound from Swansea to London with 1,780 tons of coal. The survivors were taken on board the *W&S* and landed in Newlyn at 9.00am. With enemy vessels still in the area the lifeboat then returned to sea and continued searching until 3.00pm without result.

The SS *Solstad* (1379grt) was launched in 1924 by Lewis John of Aberdeen, for Edward T Lindley, as *Gatwick*. She was transferred to Swedish ownership in 1931.

The wreck lies at 49.56.02N, 0.5. 28.13W

Three other vessels from Convoy WP457 were sunk by German E-boats, which approached from the shore side, having been in hiding in the dark with engines cut, in the bay at Porthcurnow below the Treen CG hut. It has been reported that British army lookouts failed to recognise these as enemy craft. The casualty vessels included the British naval auxiliary escort vessel HMS *Wallasea* (545gt), the UK motor vessel *Underwood* (1990gt), carrying government stores and military vehicles from Clyde to Portsmouth, and the UK-flagged steamship *Polperro* (403gt), with coal from Manchester for Penryn, which was lost with all hands so close to its destination.

John Knifton, a local historian and blogger, followed through the aftermath of this event, sharing details provided by war graves researcher David Betts. "In Penzance Cemetery lie the graves of 22 Second World War casualties from four individual ships: These vessels were in a convoy which was attacked by six E-boats as part of German attempts to disrupt the Allies' preparations for an invasion of Western Europe. The *Solstad* was torpedoed by the German torpedo boats, S-136 and S-84. The ship sank in three minutes, with the loss of five lives.

Alide Reicher, 53 years of age, was a stewardess on the *Solstad*. She was Swedish and was serving on a (neutral) Swedish merchant ship. Two British seamen, John Smyth, fireman, and Kenneth Allen were also killed. Kenneth was a deck hand from Blyth Northumberland, aged only eighteen."

His Majesty's Trawler *Wallasea* (T-345) was an Isle Class Armed Trawler built in 1943. This vessel was part of the Royal Naval Patrol Service and was 545 gross tons. Seventeen members of the Wallasea's crew are interred at Penzance, out of a total of 35 fatalities. These include an Able Seaman, the Cook, an Engineman, a Leading Steward, an Ordinary Signalman, a Seaman, a Second Hand, a Stoker and a Telegraphist.

The MV *Underwood* (1990gt) was travelling from the River Clyde to Portsmouth, with military stores including vehicles. The crew of 15 seamen and three passengers were all lost. The wreck of the *Underwood* was identified in 1975 by information on the boss of the propeller. The grave in Penzance is that of the Radio Officer, Alexander McRae, 43 years of age, from Carluke in Lanarkshire. Alexander's parents were William McRae and Annie McRae (nee Wilkie). His wife was called Edith.

The MV *Polperro* (403gt) registered in Fowey, had sailed from Manchester with a cargo of coal, bound for Penryn. The *Polperro* went down with the loss of all hands, namely eight Merchant Navy seamen and three Royal Navy gunners: Capt R.M. Hawkins, T. Clark, A.R. Cornish, D.R. Denholm, A. Georgeson, J. Georgeson, W. Hall, E. Sear, Jack Ellis. The Penzance graves from this nautical family are two Able Seamen: W. Whitelaw and J.R.W. Kirkpatrick. Jack Ellis, aged 28 and from Liverpool, had been trained as a gunner and this was his first trip. He had been married to Edith Jarvis at Walton in 1940.

The attackers on January 5th-6th 1944 were the 5th Flotilla led by Leutnant-Kommander Karl Wilhelm Walter Müller. At 3am on January 6, the British convoy was about to cross Mount's Bay where:

The squadron of E-boats (Germans called them S-Boats) was later responsible for the attack on US forces preparing an invasion force at Slapton Sands in Devon with the loss of hundreds of lives.

The weather was fine with good visibility. It was moonlight with a South-West wind Force 3 and moderate sea.' Müller had the advantage of complete surprise, coming from the landward side. The convoy's escort, led by the ageing destroyer HMS *Mackay*, was overwhelmed by the firing of no fewer than 23 torpedoes, and four ships were sunk. The German force's first attack sank the *Solstad* and the second, some five miles south of Penzance, sent the *Underwood*, the *Polperro* and the *Wallasea* to the bottom. The rest of Convoy WP457 continued on their way, while the brave crew of the Penlee lifeboat made valiant attempts to rescue any survivors. In total more than 60 people were killed.' Johnny Drew gives a gripping account of his part in the drama:

The Fowey owned and registered Polperro was built in 1937 and is seen here in 1938 at Charlestown, which was then an active port in the china clay trade
COURTESY PAUL WALKEY

"This morning I was rung up, 1.30 in the morning, by the Commander of the Coastguards, who said 'Have your pencil ready. Now be very particular what I'm telling you, mark it down and be sure to put it in practice. There's been a convoy attacked between the Runnel Stone and the Lizard. Several ships are sunk, including the escort vessel [*Wallasea*]. Quite a lot of Navy personnel in the water because it was sunk. Several ships been hit and some sunk, and their crews will be either on rafts or liferafts. As soon as you launch turn on your

"The older men couldn't do what I was doing. I was lifting the casualties off bodily

lights.' Well, this was prohibited normally, you didn't show a light. 'You know the area you have to go to?' 'Yes, sir' 'The Coxswain is also being informed. Now be careful with this, underline it. Make sure before you launch, turn on every light you've got. Searchlight, navigation lights, every light you've got. And make sure they are kept on because this convoy has been attacked by E-boats. Our Navy craft are out in pursuit of them and they won't take any chances. They will know the boat with the lights on is you, because they are informed by us. But make sure your lights are on.'

Well, in the war days, instead of firing maroons, you went from house to house, collecting the crew. So, therefore, the ones who lived nearest were called, and the Cox'n called the nearest crowd up by him. Eventually we launched. I had a bit of controversy with the Cox'n at first, because I don't think he fully understood the command he was told. He forgot about the lights, see. I said 'All lights have to go on, or we shall be blown out of the water. These were my instructions.'

So we launched on lights. We were steaming about 20 minutes and nobody seeing anything. At last, about five miles off into the South West, into the opening of the Bay, somebody spotted a dim red light. Well, they carry these flares, see. First it was two men on a raft. One was an airman and we picked him up. He had a broken leg. Wasn't long before another flare was sighted and it was a full ship's crew of 14. It was the *Solstad* of Sweden, that was her name. We picked them out with our searchlights, the rafts having these bright yellow canvas coverings. One that we picked out, all you could see was faces. The

126

water wasn't bad so we got to the raft and shouted to them. None of them could help themselves. We stopped the engines and we were drifting with the raft and brought it alongside. And all these 14 people included three women aboard. Eventually I was lifting them bodily and got the crew on board. We had a cabin in the midships you see, and those who could get down, went down there. We cut the raft adrift, and we couldn't stay much longer because these were all seriously injured. I conferred with the Cox'n that we ought to get in. We decide to radio in, and when I got to the cabin I said briefly, 'proceeding harbour, doctor and ambulance'. That's all we could say.

So, we started the engine and proceeded to Newlyn harbour. Now, it wasn't daylight yet, and Newlyn was full of Navy personnel waiting for our arrival. When we got in, the ambulance was dealing with the bad. Nothing but blood and muck all down the cabin where they got to. One or two were laid down and covered on deck on account of leg injuries. That was a very nasty turn out. As soon as we discharged them and they were gone, we were ready for sea again, because there were seven ships [in that convoy]. But as daylight comes a big Navy ship was sent. One of them picked up the 19 bodies of the escort ship and they had a communal burial at Penzance. One crew landed in Prussia Cove, a whole crew landed in there. Of course, we went out again in a similar direction but we didn't pick up any more. Well, that day was pretty hectic.

See, what it was, I had super strength given. The normal crew men were called up in the patrol service in the Navy. So we had older men, around their sixties, and perhaps one or two youngsters going on around 16. But my mate and I were two of the youngest. I was on the raft. Normally you wouldn't leave your engine, but our engines were stopped and we were drifting. Well, the older men couldn't do what I was doing. I was lifting the casualties off bodily. Three stewardesses I think, and one of them was one of the officer's wife, all injured. It

PHOTO: KNIFTON/BETTS

A total of 22 seafarers, victims of the E-boat attack on Convoy WP 457, are buried in Penzance cemetery. Seventeen of them were crew members of HMT *Wallasea*, and five from the neutral-flagged SS *Solstad*, including a deck hand and one of three stewardesses on board.

so happened that the air raid warning that had come up in the afternoon was a German reconnaissance plane which was spotted coming out of the Channel, went back and reported, and the E-boats that were stationed on the Channel Islands came up under Porthcurno and lay in wait there. As soon as they did their attack they were gone, back to the Channel Islands. That was the main convoy attack that W&S attended."

1944 SEPTEMBER On the night of **September 24/25** the lifeboat was launched to a report of three barges adrift. The lifeboat left the slip at about 20.15 and arrived at the given position 50N 0.6W, about 21.45. Cox'n Blewett reported: "I met a tug boat and asked the Captain if he had seen any craft in distress; he replied 'No'. I searched around and found a big pontoon which had broken adrift from the tug." The search was continued until 23.15, when a message was received over the Radio Telephone to return to station, arriving back at 01.15 on the 25th. In fact there were six barges lost: LCT Mark lll craft Nos: 480, 488, 491, 494, 7014 and 7015.

1944 OCTOBER An entry in the Motor Mechanic's log for **October 17** noted that the boat was launched for 'film making' at 3pm. 'Returned to slipway at 5.30pm. Petrol used 12 gallons. Boat and winch engines satisfactory.' The sea was moderate, the wind WSW.

Lighthouse sinks submarine

1944 DECEMBER Following a telephone message from the Royal Navy to the station mechanic the *W&S* launched at 11.50am on **December 18** to proceed to vicinity of the Wolf Rock Lighthouse where a German submarine had struck the Rock and sunk. The lifeboat arrived at 1.15pm and found HM Destroyers, Rescue Marine Launches (RMLs), High Speed Launches (HSLs) and aircraft in cooperation.

The Cox'n reported "We steamed through a large area of heavy oil, sighted and picked up life saving equipment, which we found to have enemy markings. We continued searching and could find nothing more. RMLs and HSLs were now returning to harbour and the Cox'n decided we could do no more so we returned to slipway and rehoused boat at 5.30pm." Naval vessels picked up 42 survivors.

The sub was U-1209, a 500-ton Type V11C U-boat. She carried a complement of 51, under the Commanding Officer, Oberleutnant zur See Hülsenbeck, who died of a heart attack on board HMCS *Montreal*. At the time of her sinking, according to the captured survivors, the U-boat was pursuing an aircraft carrier. Early that morning a heated argument had taken place on board U-1209 between Hülsenbeck and a Navigational Chief Petty Officer. The CPO reported that the U-boat was on collision course with Wolf Rock, but Hülsenbeck insisted that they were well to the west of it. At the height of their disagreement, U-1209 struck the rock.

On the Wolf Rock Lighthouse, Charles (Charlie) Cherrett was 'acting keeper

ROYAL NATIONAL LIFE-BOAT INSTITUTION.
WRECK SERVICE.
Copy of all the messages passed by the PENZANCE
PENLEE Life-Boat Station. Coastguard Station to the Life-Boat Authorities.
Name of Vessel in distress ("....") SUBMARINE UNKNOWN

Date	Time	From whom Received	To whom Sent	Means by which Messages are Sent or Received	Message
18/12/44	11-39 AM	Coastguard Penzance R.N.O. Penzance	Mechanic Penlee L/Boat	Telephone	From RNO Penzance. Launch L/Boat proceed in vicinity of Wolf Lighthouse. Submarine in distress.
---	2-45 PM	Lands End Rdo	SO Penzance Penlee L/Boat	---	Weather W1-2. Bar. seal. V1.8 Lifeboat called 1333 GMT. Have you anything for us. Reply We are still searching standing by
---	3-10 PM	---	SO Penzance	---	from Penlee L/Boat. We are returning.

REMARKS: L/Boat launched 1155. RML 534 also proceeded.
L/Boat returned 1815.
LW 1-30 P.M.

in charge'. According to an account by his grand-daughter, Lynda White, Charlie (left) and the two other keepers heard a terrifying, metallic scraping noise. When Charlie heaved himself up onto the small window space he could see a German U-boat sitting high and dry on the rock below them. His first signal to Trinity House in Penzance reported the U-boat had grounded, and the heavy swell had washed it back into the sea, only to throw it back on the rocks once again. The lighthouse then reported that the U-boat had slipped off again and was now proceeding westwards, and that a man was seen standing on the conning tower.

The U-boat had been badly holed aft and was stern heavy, with water rising in the diesel and motor rooms. Although the submarine had tried to dive, she was forced to surface but the lack of air pressure on board made it impossible to open the conning tower hatch at this stage. Submerging again, she pumped fuel oil out to enable her to float to the surface, where the Commanding Officer opened the hatch and was already on the bridge himself when he gave the order to 'Abandon Ship'. Hans George Claussen (Engineering Officer) stayed below deck to scuttle the boat and was blown out of the conning tower just before U-1209 sank. He had been the most popular officer on board, according to the interrogation notes. Although he was picked up by Allied rescue boats,

U-boats were equipped with Enigma code breaking machines. If captured the codes would be revised, which they were within days

he later died from his injuries in Penzance hospital and was buried in the town. It was later reported by the crew that while they were abandoning ship at 10.45, Hülsenbeck threw a rating off the liferaft so he could save himself. At 11.01 Charlie Cherrett sent a further signal to Trinity House stating the U-boat had sunk about one and a half miles off Wolf Rock and the crew had taken to liferafts. HMCS *Montreal* and HMCS *Ribble* were sent to pick up survivors.

There have been claims that the Allied ships torpedoed the U-boat, but in fact she sank before they had arrived on the scene, as was confirmed by the lighthouse keepers who were providing a running commentary, broadcasting over an open radio channel. Presumably with this in mind, the Admiralty made the event public knowledge, by reporting the incident to the BBC. German radio later stated that, according to an enemy radio broadcast, U-1209 had run on to the Wolf Rock and must be presumed lost. The significance to the Germans was that their U-boats were equipped with Enigma code breaking machines, which the Allies were trying to decode, and if one machine was captured the codes would need to be revised. Which they were within days.

Forty three of the crew of 51 were picked up and 42 sent as prisoners of war after interrogation. U-1209 was Hülsenbeck's first command. He was 24 years old, very inexperienced, and seems to have been at odds with his officers and men. He had forgotten to pick up his cypher books before leaving port, so he was unable to decode any cyphers intended specifically for the commanding officer. Interrogated prisoners also said that U-1209 was badly built and leaked if she dived below 120 metres. U-1209 had sailed for the North Atlantic from the Baltic. Passing to the north of Scotland she had several opportunities to engage Allied shipping but had declined, apparently to the dismay of her crew. She then made

for the west coast of Ireland, and Admiralty interrogation notes record that she tracked an aircraft carrier from Donegal to Milford Haven. According to the Admiralty there was no aircraft carrier anywhere in this area at the time.

Another witness to the event was Lieutenant (later Lt Commander) Richard Perrin of Stithians, who went to the scene in the Royal Navy's *RML 542*, operating out of the naval base at Newlyn. "I had a good crew and three of us still meet," recalled the veteran, who later had contact with some of the surviving Germans who were brought back to the Cornish port, one of whom wrote to him: 'I find it great that I have the chance, after more than 50 years, to be confronted with the events of the past and to refresh my memories and to make contact with the men who saved our lives.' While in command of *RML 542* Richard and his crew had saved many British airmen and mariners. Yet that dramatic episode will always remain vivid to him

"We were based mainly at Falmouth and carried out rescue work from there as well as from Newlyn, St Mary's in the Isles of Scilly, Appledore and, for a time, at Padstow." Perrin recalled. "On this December day we were at Newlyn, moored at the end of the North Quay. That morning I went up to the Naval Office and there was a lot of confusion going on. They had received a message from Trinity House at Penzance that there was a submarine on the Wolf. It was a million to one chance that this could happen. [Apparently] a lighthouseman had come outside that morning to spend a penny and found he was doing it on a U-boat. There was no submarine believed to be near Cornwall.

"Off we went at a rate of knots and, after passing Mousehole, we could see an aircraft carrier coming across through a 'swept channel' that had been cleared of mines, together with two destroyers. As I got nearer to the Wolf I could see part of a sub disappearing under the water and also two Canadian destroyers. One of them was dropping depth charges. The tide had come in and she had floated off. The crew tried to get a hatch open. Submariners were bobbing up in the water. We picked up some and were told to bring them to Newlyn where our intelligence corps came and took them away. The whole incident was absolutely shattering: that is why no one in the Newlyn office could believe it at first."

Three quarters of all German U-boats were sunk by Allied forces during World War Two, but this is the only known instance of a Lighthouse sinking a submarine during the war.

The U-boat sank at 49.55.46N, 05.44.30W

1945 JANUARY When off Cape Cornwall, 3 miles N of Longships, at 2.35am on **January 3, 1945**, the Cardiff steamer *Strait Fisher* capsized. Her cargo of grain had shifted, leading to the boiler room flooding and eventual capsize. A strong NW wind was blowing, with a very rough sea. The *W&S* was launched at 4.20pm and reached the position given at 6.15, but found only an empty boat. A Naval trawler had picked up 10 of the crew; three others were missing. After searching until 8 o'clock the lifeboat returned, reaching Newlyn at 9.45pm, five and a half hours after putting out. It had been an arduous service in bitter weather and heavy seas, and an increase in the usual money award on the standard scale was made to each member of the crew. Standard awards to crew and helpers, £25 0s. 6d; additional rewards to crew, £2 12s. 6d.; total rewards, £27 13s.

Three bodies had been recovered by escort vessels and two were washed up ashore. Among the dead were: Albert Henry Lesslie (58), Bosun, of Appledore; William Drummond (52) Fireman, of Glasgow; William Edwards (43) Able Seaman, of Amlwch; William Hugh Hughes (41) Master, of Amlwch; James Neil Taylor (39), Second Engineer, of Prestwick. The *Strait Fisher* (ex *Dragoon*) built 1917 (573grt) at Aberdeen was owned by James Fisher and Sons, Ltd, and registered at Barrow-in Furness. She was on passage from Plymouth to Silloth.

The wreck lies at 50.07N, 05.46W

At 3.12pm on **January 21** a telephone message from RNO ordered the lifeboat to launch at once to a ship torpedoed 180° and 4 miles from Longships The *W&S* launched about 3.30pm and proceeded to the position given and found the ship abandoned by the crew who had been taken charge of by HM ships. The Cox'n spoke to an officer of one of the ships. 'He told me the crew were all saved except two. I stood around for a while, then decided to return to station as several HM ships were dropping depth charges and their craft were in the vicinity. When I arrived at the slipway there was a snow shower and more wind so I took the lifeboat to Newlyn arriving at about 7.15pm.'

The Sennen Cove lifeboat was also tasked to go to a vessel four miles SW of the Longships. A light Northerly wind was blowing, and the sea was smooth. The Sennen lifeboat *The Newbons* put out at 3.30 and at 4.40 found that a Liberty ship, the American *George Hawley*, had been torpedoed by U-1199, with the loss of two of her 68 crew.

Survivors were rescued by the tug *TID-74* (UK) and *S. Wiley Wakeman* (USA). *George Hawley* was taken in tow by HMS *Allegiance*. All her crew except five had been taken off or got away. As everything was covered with oil, the five men got into their own boat and the lifeboat towed them to a tug, which now had a rope attached to the *George Hawley*. She got back to her station at 7.43 that evening. It was learnt later that two lives had been lost and that the *George Hawley* had been towed to Falmouth. Rewards: Sennen Cove, £24 1s.; Penlee, £20 6s.

The SS *George Hawley* was a standard EC2 Liberty ship built in 1944; keel laid April 6, launched May 20. Beached near Falmouth and refloated in 1946, then towed to Bremerhaven, loaded with obsolete chemical ammunition and scuttled in 1946.

On **March 21**, at 4.15pm, Penzance Coastguards reported that one, possibly two, vessels required help 9 miles west of the Lizard. At 5.30 pm, in thick fog, the Penlee Lifeboat found the American Liberty ship *John R. Park*, of San Francisco, sinking after being torpedoed by enemy submarine U-399. Royal Navy ships were standing by and had already rescued the whole crew of 76. Some of those had returned to the steamer to gather some of their belongings, and the *W&S* stood by while this took place. The *W&S* then returned to station, arriving at 7.30 that evening.

The news of the sinking of the *John R. Park* was given to the lifeboat

Liberty ships

During the first nine months of the War, the British merchant navy lost some 150 vessels to U-boat attacks. A programme to replace them and increase the capacity of the fleet was urgently needed and by 1940, Britain and their US ally had designed a basic cargo vessel of approx 10,000 dwt, and 135m length that could travel at over 11knots under a limited power of 2,500hp. On September 27, American President Roosevelt announced Liberty Fleet Day. Shipyards on both coasts of America started building these vessels at great speed, eventually turning them out in an average of about one month. In 1942 one ship (yard No 581) was launched after 10 days and delivered five days later. In total, 2,710 of the Liberty MCE vessels were built. At least four of them featured in services carried out by the W&S. The *George Hawley* was built at New England Shipbuilding, Portland, Maine in 1944. The *John R Park* was built in the same year at Permanente Metals Corp, Richmond, California.

station at The Lizard by the Cadgwith coastguard at 4.23pm, and at 4.56 the motor lifeboat *Duke of York* was launched. She reached the scene at 5.50pm to find HM ships and the Penlee lifeboat already there, stood by for a time, and returned to station, arriving at 8.30. Rewards: Penlee, £20 2s.; The Lizard, £25 13s.

The *Park* was the lead ship in convoy TBC102, travelling westbound from the Thames to the mouth of the Bristol Channel, but had only joined the convoy at Southampton and was heading back to the USA. She was an armed Liberty ship and as lead vessel of the convoy, she was carrying six RN personnel, the Commodore and staff. She was hit on the port side, opposite No4 hatch below the waterline. The breadth of the main deck was ruptured by the explosion. She suffered a broken propeller shaft and steering failure. Both after holds, the engine room and No3 hold were flooded. She was abandoned but did not go down immediately, and the master and 10 men reboarded, but left again when she settled down, sinking at 19.35. Five boats full of survivors were picked up by the SS *American Express*, the sixth boat was towed to Newlyn by an RAF Air Sea rescue boat.

The wreck of the John R Park lies at 50.00.04N, 05.24.35W.

A message was received at Penlee from the Coastguard at 7.20am on
March 26, that a vessel 4½ miles south-west of The Lizard needed help.
A light North-East wind was blowing. The sea was smooth. At 7.40 the
W&S was launched and on reaching the position given at 9 o'clock was
told by a Naval vessel that the Dutch motor vessel *Pacific* (362gt), from
Maryport for Penryn, part of convoy BTC108, had been torpedoed and
sunk, and that an escort vessel had rescued part of her crew. The lifeboat
searched for the remainder, but all she found was a ship's boat floating

keel up. She turned it
over to see if anyone
was underneath, but
found no one. She
took the boat in tow
and spoke to another
Naval vessel which
told her she had been
ordered to stand by,
in case survivors from
the enemy submarine,

The *Pacific* was a
362 ton general
cargo ship built in
1938 at NV Noord
Nederlandsch
Scheepswerven,
Groningen,
registered at
Groningen and
owned by C.
Tammes.
PHOTO COURTESY
ARENDNET.COM

which had been damaged by depth charges, should come to the
surface.

The lifeboat also stood by while more depth charges were dropped.
She then returned to Newlyn Harbour, arriving there at 2.15 in the
afternoon. The Lizard lifeboat was also called at 7.29am and the motor
lifeboat *Duke of York* was launched at 8.18. On reaching the scene at
9.18 she searched for two hours with air-sea rescue boats and the Penlee
lifeboat, but found only oil on the water. She returned to her station at
1.30 that afternoon. Rewards : Penlee, £8 8s. ; The Lizard, £12 8s.

The wreck lies in 45m depth at 49.54.36N, 05.17.29W
U-399, which had sunk the *John R Park* was sunk by HMS *Duckworth*.

At 7.33am on **March 29** a message was received at Sennen Cove that
a vessel needed help four miles North-North-West of Pedn-men-Dhu. A
strong South-South-West wind was blowing and the sea was rough. At
7.48 the Sennen motor lifeboat *The Newbons* was launched and at 8.25
found the corvette K458 (HMCS *Teme*). She had been hit by a German
torpedo fired from U-Boat U-315. The light frigate had lost 60 feet of her

*Dramatic photos
taken on board the
Teme of a vessel
in distress. Hand
written captions
name Lt Russ
Twining and Lt Jim
Frost securing deck
gear and ammu-
nition lockers after
the ship's stern had
been blown away*

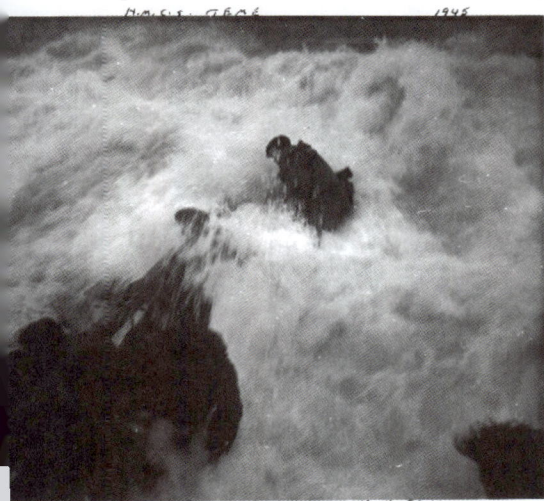

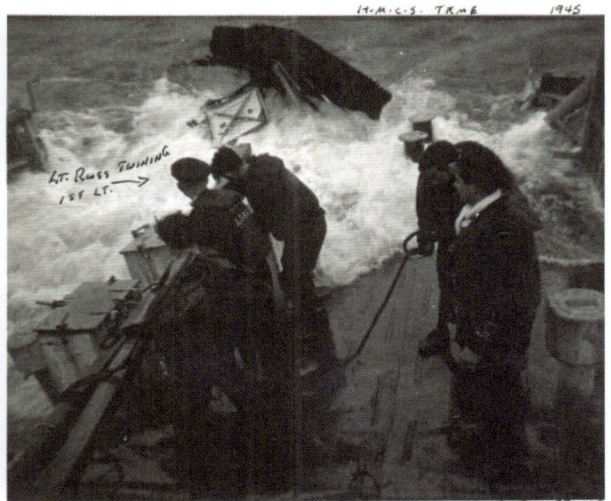

stern. The lifeboat went alongside and was asked to search in the wake of the corvette for two men who were missing. She found no one, returned to K458 and stood by until she was taken in tow by another vessel. *The Newbons* accompanied them for some way, and then, as a large tug was coming up to take over the tow, and the Penlee lifeboat *W&S* had arrived, she returned to her station, arriving at 1.15pm.

The news of a ship in distress had been received at Penlee at 8.08am, and at 8.25 the *W&S* was launched. She reached the scene at 9.45, just as a Naval vessel was connecting a tow rope to the corvette. The ships were then less than half a mile from the shore. The

The HMCS *Teme*, built as a frigate in 1943 by Smith's Dock Co Ltd, South Bank, Middlesbrough, was on loan to the Royal Canadian Navy which provided many vessels for convoy escort duties

towing started and the Penlee lifeboat stood by as well as the Sennen lifeboat, in case of need.

When they were about five miles South of the Longships Rocks, the tow-rope parted. The tug then arrived to take over the tow and the Sennen Cove lifeboat returned to her station. The new tow rope fouled the bottom and parted. The wind was freshening, the tide running stronger, and the K458 was drifting fast towards the rocks. The Penlee Coxswain advised her to anchor. The Longships Rocks were now only a mile away, and the lifeboat was asked to go alongside and take off some of the K458's crew. The crew transferred from *Teme* to a Naval tug and *Teme* was put under tow by the tug headed for Falmouth.

In the deteriorating sea conditions the Navy requested assistance from the Penlee lifeboat *W&S* to pick up the 57 crew and land them ashore. The lifeboat returned to Newlyn where she landed them and returned to her station at 5.30pm. Rewards: Penlee, £14 0s. 6d. ; Sennen Cove, £13 11s.

The landing of the crew was not counted by the RNLI as an official rescue for the *W&S* as, technically they had already been rescued by the tug, but it remains Penlee's record for largest number of casualties landed in one single service. The Coxswain's comments on the service record state 'They were kept happy with biscuits and Rum'! HMCS *Teme* was brought back to Falmouth, but was declared a total loss due to the extensive damage suffered. She was sold to be broken up for scrap on December 8, 1945.

1945 DECEMBER Red flares were observed on **December 12**, about 8 miles SSW of Coastguards at Penzer Point. The *W&S* was launched and arrived at the casualty some 14 miles from the station. They found one of HM ships had taken the Lowestoft trawler *Ala* (LT347) in tow and as their assistance was not required the lifeboat returned to Newlyn.

On **December 18**, about 11.50am, messages from the Coastguard and RAF reported a yacht dragging her anchors off Newlyn in a dangerous position. The *W&S* launched at 12.05 in SSW gales and rough sea. She found the auxiliary yacht *Diane*, crew of seven, bound from Cowes for the Mediterranean. The crew declined to leave the yacht, which had one anchor down, but had lost another. The lifeboat passed them a rope and towed them to Newlyn, arriving at 1.50pm.

Earlier that night another vessel had gone to the aid of the yacht. As then Fl/Lt Colin Bewley recalled:

"I was duty officer at RAF Air-sea Rescue Unit No 42, based at Newlyn, when at midnight we received a call from the Penlee Coastguards. I was asked to answer a distress call from a yacht lying at anchor on a lee-shore, about a mile off Penzance beach. The Penlee lifeboat, based at Mousehole, was unable to launch for reasons I couldn't learn. A gale Force SE wind was blowing, with heavy seas and a ground swell running. Sailing in *LRRC 004*, [Long Range Rescue Craft] we made our way to the yacht, which was in a dangerous position, but still holding on her anchor and cable. I was unable to get alongside, as the heavy swell and breaking seas would have rendered the attempt dangerous. Hailing her crew, I realised they were foreign, believed Greek, and our shouted exchanges were not understood. I stood by the vessel for over an hour. Finally, her crew indicated that they had no option but to rely on their anchor until daybreak, I returned to base. By next morning the weather had eased enough for the Penlee Lifeboat to tow the yacht into Newlyn Harbour. She was the Greek owned yacht *Diane* of 300 tons."

The *WMN* reported: 'After anchoring in Mount's Bay for two days, a 300-ton yacht had to be taken in tow and brought to Newlyn Harbour on Tuesday. The gale that was raging at the time had nearly destroyed her in the Bay of Biscay late last week. The yacht *Diane* is a Greek-owned vessel, and was en route to Alexandria from Cowes in the Isle of Wight. When she encountered the gale her engine broke down and her main boom was carried away. Her captain decided to let her run before the gale and endeavour to make some port in England. She was driven to the westward, and finally caught sight of Land's End, which she tried to round. The terrific seas prevented her, and she turned back and anchored with two anchors in Mount's Bay. On Tuesday her starboard anchor was carried away, and the Penlee lifeboat came out to take her in tow to the harbour. They had a hard job to get the yacht under weigh but finally they entered the harbour.'

The Motor Mechanic's Record of Services and Exercises log shows that **1946 MAY** *W&S* was taken to Ponsharden boatyard, Penryn on **May 20, 1946**, where she stayed for four months, until September 18, for what must have been an extensive refit. John Drew does not list the crew he sailed with, but the boat left Penlee slipway at 9.30am for Ponsharden Shipyard. 'Sea smooth. Wind W. Arrived at Falmouth 1.50pm (low water). Arrived Ponsharden 4pm. Turned boat over to yard. 'All machinery running exceedingly well'. Fuel: 30 gallons, Both engines @ 1200rpm. Oil temp in: 120 port and stbd: 120dg. Oil temp out: 120 port and stbd: 120dg'.

The following service was made by the relief boat *C and S* (ON690) a 45ft Watson, pictured below.

1946 JULY During thick fog on **July 1**, the Greek steamer *Gerassimos Vergottis* and the Dutch steamer *Van Ostade* had collided, and the Greek steamer had been badly damaged. The Penlee Reserve boat (ON690) and Lizard lifeboats were launched. Cox'n

Blewett could not find the casualty: 'Stopped engine, burned white flare and sounded fog horn, and searched around for three hours but found nothing.' As the boat was not equipped with radio, returned to Newlyn for more information. Learned that the vessel had steamed towards the eastern shore of Mount's Bay then her engine room had flooded and engine stopped. The vessel was picked up by a tug and beached in Falmouth Bay. Rewards: Penlee, £14 16s. 6d.; The Lizard, £20 14s. 6d. The bowman was away from home and the assistant Mechanic was not available.

1946 SEPTEMBER The *W&S* returned to Penlee from Ponsharden Shipyard on **September 18**. The sea was described as Rough, with a SW/W gale to steam into. The boat left Ponsharden at 12.20. As the weather was too bad to rehouse the boat at Penlee, she headed for Newlyn, arriving at 4.35pm. The Mechanic's log records 35 gallons of fuel consumed. 'All machinery running well'

1946 NOVEMBER On **November 17** there was a call from the Coastguard that a yacht was being driven ashore off Penzance promenade. The *W&S* arrived on scene and found the yacht *Marina* bumping heavily on the bottom, went near to yacht and found no one was on board. Then went and talked to a tug boat that was anchored in a dangerous position. He said the yacht had broken adrift from him and no crew was on board. He then asked the lb to guide him in to safer anchorage. The lb escorted the tug to a safe anchorage and a fishing boat towed the yacht to Newlyn harbour. The lb Mechanic reported that 'Could not use the [wireless] as a 7.5amp fuse had burnt out, Put in two more fuses but [to no avail . .] On return the mechanic reported the problem by telephone to Marconi at Falmouth.

1947 APRIL In **April 1947**, after a quiet winter for the lifeboat, Superintendent Cox'n Frank Blewett retired. He was the first Cox'n of the *W&S* and had served as Penlee Cox'n for a total of 27 years, and four years previously as second Cox'n. Born in 1885, he had joined the crew in 1913. His career spanned the change from pulling and sailing boats through to the first and second generation of motor boats. In his time at Penlee he is credited with saving 149 lives on the three Penlee lifeboats and also in charge of relief boats. Following the wreck of the *Taycraig* in 1936 he was awarded the Institute's Bronze medal for bravery. He was succeded as Cox'n by Eddie Madron.

HMS *Warspite*'s Final Battle

The spectacular events of **April 22/3, 1947**, and during the months following, were to became national and international news and bestowed glory on the *W&S* and her crew.

HMS *Warspite*, which ran aground in Mount's Bay, was the most famous British battleship of the 20th Century, having been active in almost all the major campaigns of the First and Second World Wars, from Jutland to Normandy. The Queen Elizabeth Class ship, commissioned in 1915, had a displacement of some 30,000 tons, and measured 643 ft x 90 ft x 33 ft. Decommissioned after WW2, and with her heavy guns removed,

she was being towed from Portsmouth to the breakers in Scotland by the Navy's most powerful tug, *Bustler*, and the two-year-old *Metinda III*, (ex-*Empire Jean*), and with a skeleton crew of eight, commanded by Captain Alan Baxter. The 'Grand Old Lady' was not prepared to go quietly, however. This was to become an ongoing drama of epic proportions. The story was eventually reported at length in The *Lifeboat* Journal of October 1947, but we refer to Johnny Drew's unpublished first-hand account of his, and 'our', boat's greatest achievement.

“It was around midnight on 22nd of April, 1947 that we were instructed to get all the crew together and stand by for instructions. Informed that the Lizard lifeboat station were on the same errand. We were there in the boathouse for about two hours, waiting till the telephone rang and said 'stand down, as the *Warspite* wasn't in any danger'. So we stood down and came to our homes. We had a powerful radio receiver at home and I took the habit of tuning in to the shipping, the distress band. When the *Warspite* got into difficulties, in our vicinity off the Wolf Rock, I was concerned that I didn't miss a message. I could hear the tugs talking, so I knew first hand that she was adrift and broken from her tugs. The weather forecast put out that it was going to increase to gale force or even severe gale force, and I don't think I did go to bed that night, I was so involved listening first hand. It then came out around midnight and the early hours of the 23rd that the tugs were going to attempt to tow to Falmouth Bay. But the weather was so bad, and she was such a huge ship that the tugs couldn't really handle her. They felt, with the force of the weather, it was best to tow her to the nearest possible safety, which was Mount's Bay, around the Lizard into Mount's Bay. They were going to attempt to pull the

HM Rescue Tug *Bustler* (111gt) *(left)* was built in 1941, with a power of 3,200hp, and *Metinda III* built 1946 had1,250hp. Eight days earlier they had helped recover the grounded liner *Queen Elizabeth* in Southampton

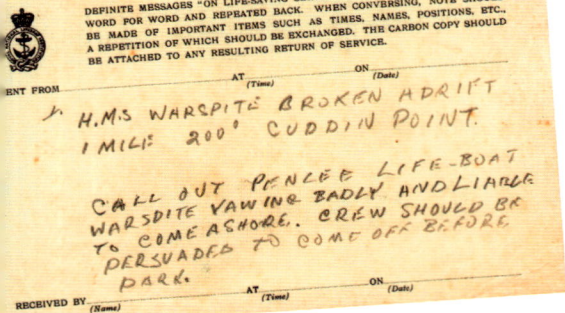

The handwritten message in the image reads:

LIFE-BOAT STATION
1959 PENLEE
1775

DEFINITE MESSAGES "ON LIFE-SAVING SERVICE" SHOULD BE WRITTEN DOWN
WORD FOR WORD AND REPEATED BACK. WHEN CONVERSING, NOTE SHOULD
BE MADE OF IMPORTANT ITEMS SUCH AS TIMES, NAMES, POSITIONS, ETC.,
A REPETITION OF WHICH SHOULD BE EXCHANGED. THE CARBON COPY SHOULD
BE ATTACHED TO ANY RESULTING RETURN OF SERVICE.

SENT FROM ____ AT ____ ON ____

H.M.S WARSPITE BROKEN ADRIFT
1 MILE 200° CUDDIN POINT.
CALL OUT PENLEE LIFE-BOAT
WARSPITE YAWING BADLY AND LIABLE
TO COME ASHORE. CREW SHOULD BE
PERSUADED TO COME OFF BEFORE
DARK.

RECEIVED BY ____ AT ____ ON ____

Warspite crab-wise, where one tug had a wire on her stern and another would have a wire on her bow. And going into this gale one tug would go this way, and the other would tow the other way. But she was always going to leeway all the time, so that's what they were doing through the night, because it was getting pretty dramatic now. And I thought she's going to go ashore in Mount's Bay. It was night time now and they eventually got her into the East side of the Bay off Mullion and Porthleven. The tugs were there, but way offshore, and they dropped anchor in that position between Mullion and Porthleven.

Now the weather was severe gale and the forecast wasn't improving; it was getting worse. When daylight came the wind didn't alter at all. It was from the South West, and the Warspite was on a deadly shore. Now, with the flood tide she had her anchor down, and the strain was so severe. Midway through the morning I went to the lifeboat house. The first thing I do is go on the radio, so I was in touch with all that was happening. I had the engines warmed up, because I felt sure we were going to be called. Then I went home and I was in the midst of having my dinner, up around one o'clock when one of the crew, second mechanic Jack Wallis, called to our back door and shouted "Johnny come on." I said "What's on?" "Warspite's adrift in the Bay."

She had broken from her anchor cable and was being driven towards Malpas Ledge. Everyone on the sea front could see a huge great ship like that. I left, got to the Front. Then right away the Coastguards fired the maroons and

"I wouldn't say it was frightening but I was a little bit scared

called out the lifeboat. We ran to the lifeboat station and we launched about 2 o'clock and proceeded towards her. But the ledge she was on, Mount Malpas Ledge, there wasn't more than a fathom of water on that ledge. The surf from the sea was churning up so much that you could almost touch the bottom with the boat's paddle. The crew had shouted at them to come off and they wouldn't. They refused, saying they were safer on the ship than on the lifeboat. The Cox'n said "Well, we can see our life is at risk." Because one nasty sea could capsize us. "We are not stopping in this position. If you're not coming off we'll return to Newlyn."

On returning to Newlyn, there was Press and radio reporters wanting news of what was happening. The crew went back to Mousehole around three or four o'clock. As was natural to me, I stayed on the lifeboat. Because in any case I had to refuel again and I made sure the tanks were full. Because I felt sure we were going to be called out again I sent one or two of the crew to have petrol sent out by car.

I remained there putting everything shipshape again. I was on the verge of coming ashore now, locked everything up, because there was nothing else to do. I was berthed about six boats off the end of the pier, the lifeboat was. After I locked up, I didn't even get to the pier, got to about the second or third boat and I could hear men shouting. There was the lifeboat crew running down over the pier and waving to me to start the engines. So I ran back and started the engines, had them running. They were all aboard now, casting off and in the harbour, putting on our oilskins and lifejackets.

When we left the harbour there was a cross sea, which means right on

The W&S shown making its approach on the port (lee) side of the battleship which was stuck fast on the rocks

the beam. The sea was South East and the wind was South Sou'West, with a full sea on the starboard beam. As we steamed further along that course there was a lot of shallow patches all along that coastline. We had to go off at a certain degree off land, and as we were approaching what we call a butting beam on our port beam, you could see the superstructure of the *Warspite* on the cliffs. There she was, and the sea was going over us, terrific sea. The Cox'n went further on to eastward until we got a full vision of the *Warspite*. She was lying right in to the sea and the wind, so that gave us no lee, what we call a leeside, to go on [for shelter]. Every 20 feet you were running into very shallow water. Well, I wouldn't say it was frightening, but I was a little bit scared, because the Cox'n altered course and ran in towards the *Warspite*. As we were getting closer in and could see these terrific seas, bow right down under the water and right up to her steering bridge and all that. The breakers were going up there because she was down by the head. The waves were breaking on the port side, but there was a channel of about 40 yards between us and the port side of the Warspite.

In discussion with the second Cox'n and three or four of us were talking in the cabin, because you couldn't stand on deck. We had to look astern to see the seas that were coming, rolling up, as we call it. If one hit you, you went up. We were were aiming for the *Warspite* at full speed, doing 8 knots. And as we were running towards her this channel was becoming more clear, the Cox'n decided he thought we would make it. I had everything battened down. "Everything alright boy?" "Everything alright Cox'n." "OK."

We didn't go more than a few yards and the sea was rolling up at the stern of us. Well, if it hit us at the speed we were going, I don't know if I would be here now. But the Cox'n saw it coming. He said to me right away "Ease the engines down". So I eased down to just below half speed. Well, that sea did break, it broke just as it hit the lifeboat. It swept everything, the lifeboat was under water and all our hatches were closed, no water could get below, but it did what we call 'swept' her. It rolled in between us and the *Warspite*. All you could see was this breaker rolling in, a great white one rolling in to the shore. Now, if we were going at any speed we would have gone in with that sea. But we slowed down. It hit us and bounced us in. Now you can see the danger? They normally come in threes, big seas like that. We had a couple more of those but, having slowed down, that gave the safety of the lifeboat. So, you put to the channel and the Cox'n said "I think we can make it."

The painting titled *St George's Day 1947. Penlee Lifeboat the W&S rescues the salvage crew from HMS Warspite, ashore in Prussia Cove*, is by the celebrated marine artist Geoffrey Huband, who was a friend of Johnny Drew and took on this epic work following Drew's recollections.

Now acknowledged for his sail-era pictures, and book covers for Alexander Kent's Richard Bolitho novels, Geoffrey began model making as a schoolboy, and he won a prize with his model of a Plymouth lifeboat. He was born in Worcestershire in 1945, and studied at Stourbridge College of Art and at Manchester University. As a student in the 1960s, he came to admire painters of the Newlyn School, who depicted day-to-day subjects in the fishing villages of Cornwall.

Geoffrey visited Mousehole on family holidays and started painting inside the Penlee boathouse, where he met Johnny, who took the young boy under his wing. Geoff remembers him as simple, straightforward and very proud man. 'You could always tell what he was thinking. He was very fastidious.

"One of my earliest childhood memories is of a breezy day on a Cornish clifftop, when my father lifted me up to glimpse a scene of great local interest. It was the Battleship HMS *Warspite*, run aground on the rocks of Prussia Cove. I was about two and a half year's old at the time. Johnny's account of the rescue inspired this painting, which is a tribute to him and the extraordinary bravery of the crew of the W&S."

So we're going now, not dead slow, but steering our way on the lifeboat. We went down this channel, nothing but white water, just steerage way. When we got down towards the port quarter [of the battleship] there under the stern of the lifeboat there was nothing but rocks breaking all around, so near the stern of the lifeboat. If you made a slight error the lifeboat would have touched those rocks and that would have been it. But we were relying on seamanship and entirely on machinery. They were our salvation. When we got under the stern of the ship and brought the lifeboat around, head in to the winds and sea, we got the lines onto the ship, one for the bow, one for the quarter. With the wind and the sea coming in, the second Cox'n, the second mechanic and myself we couldn't stand up, so the best thing to do was to kneel under the canopy to be steadier and have better control. Then with the Cox'n's orders for 35 minutes it was nothing but 'Full Ahead: Full Astern'. As that breaking sea was coming and hitting the lifeboat, we'd come full astern, then ahead to keep her in position. So we were see-sawing to keep the lifeboat in position. At the same time we were sometimes up level with the deck and sometimes 30 feet down. Each man jumped onto our deck, one at a time, until we took them all off. I know one man did slip and went down between the lifeboat and the ship. But a couple of the good crew members on board soon had him. We saved the eight of them.

Now, when we came down one time we could hear splinters. As the lifeboat dropped on this occasion she dropped on the Warspite's torpedo blisters. When all big battleships were built, their armour was sometimes a foot thick. Torpedoes would be set at the waterline. These torpedo blisters

[officially known as the waterline belt] were all the length of the ship, practically, just below the waterline. They would be projecting out, so if a torpedo hit, that would cushion the blow before it hit the ship. That's the torpedo blister. When we came down on that we heard splinters. We said the lifeboat was damaged somewhere under the water. We cut the ropes clear and we did not come out full ahead. The Cox'n was very cool and collected and said "Just less than half speed so we can dodge out into the weather."

So we dodged out at that speed until we got well clear of the *Warspite*, out into the open sea again. From then on things improved, not the weather, but as far as our safety was concerned, following the open seaway. Well, the Cox'n had control and could head up for Newlyn. We arrived sometime after seven I should think. Newlyn pier was black with people, thousands of them. In the fields watching, the lot.

The first thing I did then, I said to the Cox'n that we try all the hand pumps in all the compartments to see if that blow we heard in the splinter had damaged her under water. If any of the compartments was damaged it would fill with water. Fortunately there was no water. So we hadn't damaged the hull. A day or two later when the boat was rehoused on the slipway we found what the damage was. It was the handrail which is around the bilge keels of all lifeboats. So if anyone was in the water they could grab it. Two on the port and two on starboard. If anyone wanted to be rescued he had something to grab onto. Well the screeching of the wood that we heard on that occasion was one of these handrails damaged, split right off. But the lifeboat hull wasn't damaged and we didn't make no water. [Johnny denied reports that there was a problem

141

H·M·S WARSPITE
AT PRUSSIA COVE, CORNWALL

with one of 'his' engines, saying they had served him well.]

There was a ship's lifeboat slung out on a davit on the *Warspite*'s port quarter, so it was exceeding difficult for the Cox'n to bring the lifeboat into a correct position to be clear of that boat. She was an obstruction regarding the rescue. What he did was, as soon as we got in that position, we had our two masts up and our aerials were set from mast to mast. But, so that he could manoeuvre and get in that position we lowered down the forward mast onto the lifeboat, so that all two thirds of our boat would be clear, without any obstruction. So the mast was down while you did the rescue. Well, the Cox'n relied on me; he knew it all but just needed reminding that our radio communications were out because the aerials were down. When were leaving the *Warspite* to come out, there's the lifeboat's recognition to the Coastguards, and vice versa — Coastguards to lifeboat. You had recognition hand signals in practice ever since the lifeboats were built. Hand flares, and you had to memorise them, know them very well. They were your contact from shore to lifeboat, before radio was involved. So as soon as we took the crew off, and me knowing we couldn't call up on the radio and give information, I said to the Cox'n "Put out our hand flares?". "Right" he said. Took out a flare which burns green to white. When we took it out and one of the crew lit it up, it burned a second or two green and then it turned to white. That indicated to them on shore that we had taken off all the crew. Had we shown green and the other was green that meant we took off part or none. The third one, which we hope we never used, was red, if we were in trouble. But the green to white was essential. Although we were out of radio contact we had done the service and were returning. As soon as we got out into the open seaway I said to the Cox'n 'We ought to send a radio message.' 'Well,' he said 'slow her down.' So we slowed her down, to put the mast up and set the aerials. I then went down to the cabin and called up Land's End Radio and gave them the final message that we had taken the eight crew off and were returning to Newlyn.

The assistant mechanic was Jack Wallis. We had a pretty good understanding. He was very thorough, very thorough. He was a good seaman as well. When the Cox'n said 'Full ahead', it was full speed. We opened our two main throttles to full. The two throttles were open and the engines were running on the full governors. When we came ahead, it was full ahead; when we came full astern it was full astern. It was instant; it had to be. That's why the service under those conditions proved so successful. **"**

The Institution made the following awards: to Coxswain Edwin F. Madron, the Silver medal for gallantry, with a copy of the vote inscribed on vellum. To John B. Drew, the Motor Mechanic, the Bronze medal for gallantry with a copy of the vote inscribed on vellum. To each other member of the crew, Joseph J. Madron, Second Coxswain, John T. Worth, Bowman, J. C. Wallis, Assistant Motor Mechanic, and A. Madron, B. Jeffery and C Williams, Lifeboatmen, the thanks of the Institution inscribed on vellum. To the Coxswain and each member of the crew a reward of £4 in addition to the ordinary scale rewards for the assembly and two launches, of £2 8s. Standard rewards, £34 17s.; additional rewards, £32. Total rewards £66 17s.

The crew receive their Vellums from the Mayor of Penzance (Ald J.T. Trezise) at the boathouse several months after the event. The Mayor said it was a privilige to be associated with such an award to these "heroes in plain clothes"

After the *Warspite* shout on April 23/4 the *W&S* did not go straight to refit, unless for a very short visit. Presumably the splintered grab rail was repaired at the boathouse. There were no further shouts that year but there was a routine exercise on May 12, and another exercise on July 26, with DE Osbourne on board. 'Tested reverse gears and all lights. Tested radio with Land's End Radio which proved satisfactory. All machinery running well.'

On **August 8** another exercise took place from 11am -12 noon. The same **1947 AUGUST** items tested, with radio 'considerable improvement'. At 1.30pm that afternoon the boat left Mousehole for Porthleven. and from 2.45pm to 5pm the *W&S* ran trips for the public during Porthleven's annual Lifeboat Day. 'Very satisfactory radio tests with Tenby Coastguards, Sevenstones and Land's End'. Left at 6.45, rehoused at 8.15pm. There were further exercises in October, November and December 1947.

Helicopter relieves lighthouse keepers

The winter of 1947/48 was one of the most severe to hit Britain for years. **1948 FEBRUARY** Owing to the prolonged bad weather at the end of January and the beginning of February, it had been impossible to relieve the keepers of the Wolf Rock Lighthouse by sea, and the relief was so long overdue they had run very short of food. Trinity House arranged for a civilian helicopter to fly out from Westland Helicopters' base at Yeovil to drop 200lb of supplies on **February 7, 1948**. This was the first time such an operation had been attempted by helicopter. As there was a 40mph wind blowing from the SouthWest, and the sea was rough, the *W&S* was launched at 7.35am

DEFINITE MESSAGES "ON LIFE-SAVING SERVICE" SHOULD BE WRITTEN DOWN
WORD FOR WORD AND REPEATED BACK. WHEN CONVERSING, NOTE SHOULD
BE MADE OF IMPORTANT ITEMS SUCH AS TIMES, NAMES, POSITIONS, ETC.,
A REPETITION OF WHICH SHOULD BE EXCHANGED. THE CARBON COPY SHOULD
BE ATTACHED TO ANY RESULTING RETURN OF SERVICE.

SENT FROM Coxswain Madron AT (Time) Sec. Penlee Life Bo'd't. (Date)
Feb. 6th 48. 9 P.M. Received message from Pilot of Helicopter to.
assemble crew at 6.30 A.M. on the 7th to proceed to Wolf Rock.
further stand by him in delivering stores to Light House. + wait
instructions. At 6.30. Crew Assembled + at 7.15 A.M. Phone
message from Pilot asking if we could go to Light House
as wind was blowing about 40 miles Pr. + sea rough.
I told him it all depended on him as we were waiting. he
gave order to carry on + best of Luck. Launched at once
+ proceeded to Wolf Rock at 7.30. Strong wind + heavy Sea

RECEIVED BY (Name)   AT (Time)   ON (Date)

and went out to the lighthouse to stand by in case of accident. The helicopter lowered three bags of supplies to the lighthouse, and flew back, and the lifeboat returned to her station arriving at noon. The mechanic's logbook recorded that 'Representative of the Press and a photographer were on board'.

Newspapers reported that as many as five aircraft carrying press and newsreel teams covered the operation, although there is very little material remaining. Cox'n Madron noted: 'Blowing a gale and heavy sea. Took one extra man in case needed at Wolf Rock Lighthouse.' That made a total crew of eight; the extra man was E.F. Madron Jr, the fourth of that family on board. The Westland-Sikorsky S-51 was piloted by Capt Alan Bristow who told the Press:

"By 8.24 in the morning of the seventh we were airborne with the provisions. The route was over land from Culdrose to the Coast Guard Station at Land's End and from there over eight miles of sea directly to the Wolf Rock Lighthouse. We made the sea crossing at about 100 feet and approached the lighthouse on the north side at about the same height, the weather at this time being wind 34 to 40 knots, gusting to 45 knots, with very turbulent air at the lighthouse."

In his book *Helicopter Pilot, The Autobiography* Alan Bristow elaborated further: "Westland's chief service engineer Les Swain and I loaded the helicopter at Yeovil for the flight to Culdrose where we would refuel. Looking around for things I ought to take, my eyes fell on a pair of bolt croppers. They might come in handy — 'Better take those with us', I said to Les. Sensing a story, the Daily Express sent down a Dragon Rapide aircraft with [photographer] Walter Bellamy aboard.

"The next day I was refused permission to fly by the CO at Culdrose as the storm raged unabated. After two days of continuous gales I could stand it no longer. I wasn't in the Navy any more — I could fly when I thought it was safe to do so. On the third day Les and I

TRINITY HOUSE

lifted off at first light and set course for the lighthouse eight miles off Land's End, with a 60 knot gale on the nose. Severe turbulence made it difficult to hold position over the 18-inch-wide lantern gallery as Les lowered the three bundles of supplies with the rescue hoist.

The Westland-Sikorsky S-51 was the first commercial helicopter to come into service, piloted by Capt Bristow, a pioneer and adventurer who was one of the Fleet Air Arm's first helicopter pilots. He went on to form Bristow Helicopters, which took over SAR duties for the MCA in 2014, replacing the Royal Navy choppers from RNAS Culdrose

"The first two sacks were quickly taken off the hoist cable by the grateful lighthouse keepers, but when Les lowered the third sack, for some reason that has never been explained, one of the keepers clipped the hook to the gallery rail. In an instant the helicopter became almost uncontrollable, and as I struggled to stave off disaster with the front wheel bouncing off the lantern roof I screamed at Les 'Cut the fucking cable!' Les grabbed the croppers and chopped through the wire and, freed from the rail, the S-51 soared upwards away from the Wolf Rock like a champagne cork. Back at Culdrose, the CO was livid and gave me a severe reprimand for disobeying his order to fly, claiming I had endangered life. He was probably right. The Royal Aero Club awarded Les and me their Silver Medal for Valour."

If Bristow had not taken the bolt croppers with him that would have made a dramatic shout for the lifeboat standing by in the waters beneath. Speaking later to The Cornishman, lighthouse keeper Clifford Wheeler said "The Penlee lifeboat passed us a message that an attempt would be made to drop food by helicopter. When we first saw it it was so low we thought it was a boat. I can never forget those three or four thrilling minutes when it hovered above us. Mr John Mudge caught the line and held on to it tightly. He didn't let go even when he was hoisted three feet off the gallery. We wanted to make sure of that food." That probably explains who clipped the chopper to the rail and why.

Some 20 years later the Wolf Rock became the first rock lighthouse in the world to have a helideck built on top of the lantern housing. In 1998, the lighthouse was de-manned and automated.

At 12.40pm on **June 7**, Porthleven Coastguard telephoned that a man **1948 JUNE** was in the sea, shouting for help, 800 yards east of Porthleven pier. The W&S was launched at 12.45. A moderate Southerly breeze was blowing and the sea was rough. The lifeboat found the bodies of a man and a woman about 400 yards from the pier. The crew applied artificial respiration, but all attempts to revive them failed. Doctor and police were waiting on the pier. The doctor came on board and said life was extinct. It transpired that they were a newly-married couple who had just arrived in Cornwall on honeymoon. Their bodies were left in care of the local police. Owing to the bad weather the lifeboat returned to Newlyn, as there was too much sea to put the boat back on the slip, arriving at 4.30, and returned to her station the next day. Rewards, £15 14s. 6d.

The inquest heard that Mr. and Mrs. Charles Haddon Turner, who were drowned just three days after their marriage, had been warned by the proprietor of their hotel not to bathe there when there was a sea running. Coastguard Station Officer Alfred John Husband said that he and another coastguard fired a rocket, but it fell 50 yards short of Turner, who was 100 yards from shore. He called the Penlee lifeboat, whose Coxswain Edwin Madron, said it took about an hour to cover the nine miles from the lifeboat station.

A family tragedy that shook the community

1948 JUNE Less than three weeks later, the fishing port of Porthleven was struck by an even more tragic event that took place during the annual Sunday school festival of St. Peterstide, when seven men went out fishing and only one returned. At 11.05pm on the night of **June 25**, during a thick fog, the Cadgwith Coastguard telephoned the Cadgwith lifeboat station that the American steamer *Chrysanthy Star* had radioed that she had collided with a small vessel 10 miles South-South-East of Lizard Point. The Cadgwith lifeboat *Guide of Dunkirk* was launched at 11.25, in a moderate North-Westerly breeze with a calm sea. Another message stated the 25-ton fishing vessel *Energetic* (PZ114) of Porthleven, had been sunk, and that the steamer had picked up two of her crew of seven.

The Coastguard also informed the Penlee station, at 12.55am, and at 1.20 the *W&S* was launched. The two lifeboats made a very wide search, but found no trace of the five missing men and returned to their stations, Cadgwith arriving at 6.15am on the 26th, and the Penlee boat returning at 8.30am. An aircraft that had joined the search was recalled because of the fog.

Chrysanthy Star was built in 1944 as a tanker Raccoon, then J. C. W. Beckham in 1946, Chrysanthy Star in 1948. Converted to dry cargo ship Jupiter in 1949, Searanger in 1951, Sariza in 1953, Sara in 1963, Asia Mariner in 1965. Scrapped in 1968.

In the meantime St. Mary's (IoS) station was informed at 7.20am that the *Chrysanthy Star* had radioed that one of the rescued men had died, and had asked the lifeboat to bring them ashore, while she diverted from her course to Curacao. The St Mary's lifeboat *Cunard* (ON728) was launched at 7.46, and found the steamer eight miles South of Wolf Rock to land the survivor and the body. Rewards: Cadgwith, £27 8s.; Penlee, £18 5s.; St. Mary's, £16 12s.

Five of the six men lost that night were the Richards brothers John Henry, William, Perkin, Gilbert and Thomas. A sixth brother, Ralph, was the only survivor. The seventh man was named as Roy Mewton, a friend who had joined them for a fishing trip. Ralph and Mr Mewton were picked up by the steamer. The search for survivors resumed the next day with the destroyer HMS *Cockade,* and a Lancaster aircraft spotted wreckage but no sign of survivors. The well-known Richards family had been fishermen for generations. The brothers were all married with children and most of them sang in the church choir. The sole survivor Ralph Richards, a non-swimmer, described what happened:

"At a quarter past six on a lovely summer's evening, we left our little home port of Porthleven, after having bid farewell to our loved ones, and telling them to expect us back tomorrow at midday. We were in company with four or five other boats which comprised the long-line fleet. As we left the harbour we could see a bank of fog and after an hour we entered into this dense fog, and an hour and half longer we discussed the advisability of shooting our nets immediately or whether to wait to see if the fog would lift. At this time we were all conscious of our danger, and were sounding our fog horn at frequent intervals; then we put on our lights and were all on deck waiting for darkness to fall so that we could pull in our nets. We had heard two or three steamers pass down some distance from us, but now we could hear one approaching from the South East and, by the sound of its fog horn, we had the feeling that it was coming towards us, so we lit a flare and continuously sounded the fog horn. Not being under power, we were helpless to do anything more, but still the ship came on and on, and at last we saw her break through the fog about 300 yards from us, and coming straight for us.

Three of us went forward in the bows and the rest of us stayed aft. I never saw them again. All we could do now was wait for the moment of impact. The suspense was terrible and I can see it all happening now. Crash! Into our side went the steamer, its bows going in about a third of the way and pushing us down on an even keel. The next moment I was going down under the water, seemingly for ages, being drawn down and down by the suction from the boat.

When I came to the surface I noticed the steamer had not yet passed by, and the first thought that entered my head was the danger from the ship's propellers. So, hanging on to the float, I did my utmost to kick myself away from the ship's side. I remember at this time being conscious of one of my brothers being close at hand fighting for his life, but only for a moment, for the sea was in a turmoil and he was soon dragged down, never to appear above the surface again. I realised within myself that they were all gone and I was the only one left.

As the full force of this broke upon me, I was overwhelmed and sorely

"I heard a voice behind me saying 'Hold on old timer, we're coming'

tempted to let go; it seemed far easier for me to die rather than to live. But the Lord brought before me a vision of my wife and two dear children and I pictured all that my loss would mean to them, and so I clung fast. Three times I was sorely tempted to let go, but each time the Lord brought the same vision before me. After some time, as I looked towards the west, I saw the masthead lights of a steamer. Not realising that it was the same ship that had collided with us, I commenced shouting, 'Help! Help! Help!'

After some little time I heard a voice directly behind me saying 'Hold on old timer, we are coming!' The next moment I was taken aboard the ship's lifeboat and, as I felt someone cutting away my clothes, I became unconscious. I regained consciousness to find myself in the ship's hospital being forced to drink hot milk and coffee and being given a continual renewal of hot blankets. Another stretcher was wheeled into the sick bay and upon it lay Mr Mewton. He was unconscious but still alive, and the second mate commenced artificial respiration for about five hours. Then one of the crew told me they would like to take me to another room. I knew the reason for this; Mr Mewton had gone beyond all human aid. You can imagine something of what I passed through in the ship all night and well into the next day, wondering how, when and where I would get ashore and how I was going to face my brothers' widows and fatherless children, and my poor aged father. Continually I cried to God to see me through. **"**

The full statement can be read at www.porthlevengigclub.com/history

The *Energetic* (PZ114) was a wooden-hulled 25ton, drifter/long liner, built in the early 1920s at Porthleven: Reputedly the first fishing boat in Mount's Bay to be fitted with an internal combustion engine (diesel)

R.N.L.I.

LIFE-BOAT SERVICE REPORT.

Life-boat Station **PENLEE**   Date **Nov 1st 48.**

Name of Life-boat **W & S**   and No.

Name and Port of Vessel

Please give here an ACCOUNT OF THE SERVICE from the time of receiving first news to Life-boat's return to Station.

(If convenient the account may be typewritten on a separate sheet).

Called by Coastguards by Phone approx. 1.20 A.M. reporting lights seen from ship ½ mile off Penberth Cove. Coastguards from Treen were then proceeding to beach to find out the trouble. Approx 2.40 A.M. called by Coastguards + informed they had found one man on beach from Ship ashore off Penberth told Coastguard to put off Marconi's. Launched Life-Boat 2.50 A.M. arrived off Penberth 3.50. Searched area with Searchlight but no sign of anything about 4.30 A.M. received wireless message from Landsend Radio to return to Station Arrived at Newlyn. as weather to bad to return to slip about 5.30 A.M. Put Lifeboat ready for service 6. A.M.

NOTE.—It is particularly requested that all questions be answered and full information given.

QUESTIONS.	ANSWERS.
1—Type, Name, and Port of Vessel ?	1 French Oil Tanker.
2—Names of Master, and Owner ?	2
3—Number of Persons on board ?	3 Twelve.
4—Tonnage, and whether vessel loaded, in ballast, or how occupied, where from and whither bound ?	4 Ballast. 500 Tons.
5—Position of casualty ?	5 Gribba Point.
6—Nature of casualty ? Did vessel become total wreck? If not, state what became of her ?	6 Total Wreck.
7—Direction and force of wind ? (a) At launching place (b) At scene of casualty	7 S.W.
8—Condition of sea ? (a) At launching place (b) At scene of casualty	8 Rough.
9—Condition of weather ?	9 Fog
10—State of Tide at time of despatching Life-boat ?	10 Half Flood.
11—Did the Life-boat authority and the Coastguard communicate (Regulation 43) ? If so, attach Coastguard message form or forward it when received ?	11
12—Was telephone or telegraph used ? If so, attach Life-boat telephone pad form containing the message, and those containing any other messages	12
13—Were the adjacent stations informed, by the Coastguard or by the Life-boat Authority, of the action being taken ? (Normally the responsibility of the Coastguard)	13
14—Which Life-boat Official authorised the despatch of the Life-boat ?	14 D. Officer Coastguard.
15—Time when signal was first seen or warning received by the Life-boat authorities and from whom received ?	15 Coast Guard.
16—By whom was assembly signal made and at what time ?	16 Coast Guard.

(Continue on next page).

Maroons were fired by Coastguards at Penzer on **August 4**, after the reported sighting of two paratroopers dropped from a plane about two miles SE of Mullion Island. A message from Land's End Radio called the lb to proceed to an area two miles SW of Mullion and not SE. Another message told them to return to station as the plane had dropped kites, not parachutists.

In the early hours of **November 1** Treen Coastguard had seen through the heavy rain the lights of a vessel heading straight for the beach at Pedn Vounder and signalled it with the Aldis lamp. The vessel turned away to the east around Logan Rock headland but was never seen again. A Lifesaving Apparatus (LSA) search party reported the wind was thick with the stench of fuel oil, and in Penberth Cove they picked up one survivor found floating in the surf and covered in oil, and carried him to a nearby cottage. They also recovered a lifebuoy marked 'St Guénolé-Rouen'. The *W&S* was launched at 2.50am to reports of a vessel needing help half a mile South of Penberth. The Coastguard at Penzer reported at 4.01am that the 500-ton French-flagged motor tanker S*aint Guénolé* of Rouen, was ashore, bottom up, at Penberth Cove, probably a total wreck. *Lloyd's List* noted the tanker was bound from Nantes for Irvine in ballast. 'She is wedged between rocks close to Penberth. Little chance of salvage.' The *W&S* searched throughout the night but returned to Newlyn at 6.20am without finding any further survivors from the crew of 12. The sole survivor was 23-year-old Andre Fourecin, who had been asleep in his cabin, dived into the surf and was washed ashore to be pulled to safety. An intensified search was made at daybreak by Treen and Tol-Pedn companies without result. The master was named as Captain Dagorne.

The wreck broke up at 50.02.36N, 05.37.42W

The MV *Saint-Guénolé* was built in 1938 at Rotterdam as MV *Lind* for Norwegian owners. In 1940 blockaded in Gothenburg, and taken to England in 1942. Of 10 Norwegian ships helped by the RN to escape six were lost to German forces. In 1947 sold to Goudron of France and renamed *Saint-Guénolé*.

1948 NOVEMBER The Penlee boat was called out by Penzer Coastguards at 04.40 **November 11**, on a service to the French fishing vessel *Sapphire*. The crew was assembled by the Mechanic, who was informed at 05.25 that the Sennen Cove boat had been tasked.

Mr Barrie Bennetts, Hon. Secretary, was appointed MBE in the New Year's Honours List. He had been Hon. Sec. since 1913; was awarded binoculars in 1925; Gold Badge 1948

JC 'Jack' Wallis (right) retired after 11 years as assistant Mechanic and 19 years as a member of crew.

EDWINA REYNOLDS COLLECTION

1949 JANUARY On **January 30, 1949** orders were received from the Chief Inspector (RNLI) for the *W&S* to tow the Reserve lifeboat *H.F. Bailey* (ON670) from Newlyn to Falmouth. The boat had suffered engine trouble. Cox'n Madron got in touch with Cox'n Hill and told him that: 'Conditions were suitable as the flood tide would help us round the Lizard Head and we may be a long time before having another chance. He told me he was ready to go at once — time approximately 3pm. I told Coastguard to fire maroons and assemble the crew. Maroons fired about 3.15. Boat launched about 3.25pm. Proceeded to Newlyn to collect the lifeboat for towing and left for Falmouth. Arrived Falmouth about 8.30pm and moored Reserve lifeboat to Falmouth lifeboat. Stayed at Falmouth until 10pm, for ebb tide to go around the Lizard, arriving at the station at 2.45am on Jan 31st. Weather too bad to rehouse. Went to Newlyn 3am and stayed until daylight. Rehoused lifeboat at 9.30am.'

1949 MARCH There was a message from the Coastguard at Penzer on **March 23,** that an aeroplane from RN Air Station Culdrose, had dropped into the sea. Position approx 3 miles WSW of Porthleven. CG fired maroons. Launched lifeboat approx 14.22. At 14.25 received message from Land's End Radio that its position was now 1 mile South of Porthleven. At 14.44 received 'Mayday' message to all shipping: 'Plane down in position 1 mile South of Porthleven'. At approx 15.00 received message from Land's End Radio to return to station as the pilot had been rescued.

1949 SEPTEMBER On **September 4** a new HM Customs motor launch, *Badger*, being delivered from Bideford to Weymouth with a crew of eight, developed reversing gear trouble about 4 miles SE of Mousehole and was towed into Newlyn in the evening by the French motor crabber *Notre Salut* of Camaret. Penlee lb launched and escorted the crabber and tow into Newlyn.

At about 3.18pm on **September 19** the Penzer Point Coastguard telephoned to say that Culdrose Air Station had reported a crashed Firefly aeroplane, 3 miles South of Penzer Point. A Sea Otter rescue seaplane had landed near her and could not take off again. The Captain of the Naval Air Station requested the lifeboat to stand by. The *W&S* was launched at 3.40pm in a choppy sea, with a light South-Easterly breeze blowing, and stood by the seaplane for 90 minutes as she taxied across Mount's Bay. Five miles west of The Lizard an RAF

rescue launch took her in tow and headed for Helford River. The services of the lifeboat were, therefore, no longer required. The *W&S* was recalled to her station, arriving at 6 o'clock in the evening. The pilot of the Firefly was rescued by a trawler. Rewards, £10 1s.

Following a phone call from Penzance Coastguard the *W&S* was launched on **February 25, 1950** after the German tug *Rechtenfleth* had reported at 4.35am that her tow, the 4,000 ton German steamer *Venus* bound from Glasgow for Kiel, had broken adrift, and her own engine had broken down about 5 miles SE of Wolf Rock. The *Venus* was drifting towards the shore in a heavy swell. Land's End Radio broadcast an SOS and the *W&S* launched at 5.25am. At 5.51am the tug reported her engine working again and re-connected with the tow at 7.25am. When the Penlee lifeboat arrived the connection had been made. Two more tugs, including the salvage tug *Turmoil*, arrived on scene but were not required. *Venus* had a running crew of eight. The lifeboat returned to station, arriving at Newlyn at 9.10am.

The Turmoil *was a sister to the* Bustler *and was later to become world famous when saving the captain of the* Flying Enterprise *which sank in the Channel in 1953*

The *Western Morning News* printed this letter following the visit of the Duchess of Kent, on May 8, 1950.

Dear Mr Bennetts, I am desired by the Duchess of Kent to write and thank you for the admirable arrangements which you made in connection with the Duchess of Kent's visit to the lifeboat station at Penlee on Monday, the 8th May. Her Royal Highness was very pleased to have had this opportunity of meeting you, and was particularly glad that she had been able to have a few words with the very gallant members of the lifeboat crew. He Royal Highness had heard much of their bravery, and was so pleased that she was able to see them launch their craft, and the admirable manner in which they set about it.

Would you convey to all who are concerned with the lifeboat station at Penlee the Duchess of Kent's warm appreciation of this very happy visit which she so much enjoyed. Marlborough House, S.W.

1950 JULY The next five shouts were attributed to the relief boat *M.O.Y.E.* (ON695) The *W&S*, now almost 20 years old, was away for maintenance/repair for several months, but there is no available record of how long she was off station. Two services were made in July and November to vessels working on the salvage of the *Warspite*. These included the salvage boat *Barnet*, and the tugs *Freebooter, Masterman, Tradesman* and *Superman*.

July 15, 1950. Coastguard telephoned to say the Salvage vessel/ trawler *Barnet* was drifting ashore about 1 mile East of Penzance lighthouse. Lb ON695 launched approx 2pm and found the trawler ashore with a heavy sea breaking. There was no sign of the crew on board. The lb put head to sea and steamed offshore to find the crew aboard the tug *Freebooter*. The lb took off the five crew members in a heavy sea, and landed them at Newlyn.

July 26, 1950. A Meteor jet fighter from RNAS Culdrose crashed into the sea. The relief lifeboat was launched at 3pm but after searching in company with other vessels nothing was found except oil on the surface.

1950 OCTOBER

October 3. Another plane from Culdrose was reported ditched two miles South of Porthleven. The lb launched at approx 1.30pm, with Land's End Radio broadcasting a Mayday signal. Nothing was found and the lb's radio receiver failed. Returned to station about 4.15pm

October 9. Yet another plane was reported by Culdrose to have ditched in the sea, 15 miles East of St Martins, IoS, with the pilot reported to be in a rubber dinghy. The lb launched at approx 12.40 and arrived on position approx two hours later, in sight of St Mary's lb, one HM Destroyer and several other vessels. Searched until dark and abandoned at 6.30pm

The Warspite drama continued, as tugs involved in the salvage operation managed to get themselves in trouble
PENLEE HOUSE MUSEUM

1950 NOVEMBER

November 11. The tug *Masterman* of Falmouth, was anchored in Mount's Bay, slowly sinking after striking Hogus Rocks near Marazion, while working on the *Warspite* salvage. At 1.15pm ON695 was launched in a full SW gale and heavy seas. Stood by until the *Barnet* took her in tow. *Tradesman* caught a wire rope on her prop while trying to help and damaged her stern tube. While being towed into Newlyn, *Masterman* struck lighthouse pier, and was later repaired with a cement box. *Tradesman* was towed to Falmouth by the tug *Superman*.

The W&S was back on service on **January 27, 1951**, when Penzer Point
CG phoned at 6.25pm with information that the commander of the cable
ship Ariel (owned by the General Post Office) was sick. The ship was lying
half a mile off Newlyn Pier. Two doctors were on board but the weather
was too bad for a motor trawler to come alongside. The W&S launched
at 6.37pm in a heavy sea and strong SSE breeze, and reached the ship
about 7pm. With a heavy sea running, the lb asked for the sea ladder
and lights to be placed on the port side. The patient and two doctors were
taken to Newlyn where an ambulance was waiting. There was too much
sea to house the boat, which was rehoused the next day.

*This was the first of the 'Medico' shouts under a new arrangement
between Land's End Radio, the GPO (General Post Office, owners of the
Ariel), and local medical services, which ensured that a doctor would
be on standby to attend ship-board casualties. Medical evacuations of
this kind (generally known as medevacs) were some of the most frequent
calls for the Penlee lifeboat during the 1950s, as vessels arrived at the
mouth of the English Channel from all corners of the globe with all kinds
of medical conditions, ranging from industrial accident victims to heart
attack and appendicitis sufferers. (See the shout on December 1956.)

Land's End Radio reported on **March 16, 1951,** that a coastal vessel was
showing distress signals and in danger of drifting ashore approx 5 miles
SW of Penzer Point. In weather that
was 'thick and foggy' the W&S was
launched at 12.20 and proceeded to
the position given. At approx 13.00hr
a message was received stating the
tug Turmoil was also heading to the
scene. With very poor visibility the
lifeboat began searching around the
area. A further message informed
the lifeboat that the British coaster
Themston had safely anchored in
Mount's Bay. The boat returned to
station and was housed at 14.40 hrs.

The Themston was
a 200ft single screw
steamer of 711gt,
built by J Fullerton,
Paisley, in 1904 as
Dublin (ex-Cardigan
Coast,), renamed
Themston in 1940.
She was broken up
at Rosyth in 1952
at a ripe old age

Out of their element

During World War 2, there were, of course, frequent shouts to search
for missing aircraft mostly covered by HM forces but, surprisingly,
there was also a large number of downed aircraft in the late 1940s and
Fifties. The Royal Naval Air station at Culdrose on the Lizard peninsular
was officially opened in 1947, to be used a training facility for several
specialist squadrons of the Fleet Air Arm (Royal Navy), flying a wide range
of aircraft. In 1949, the familiar sound of piston-engine planes was often
drowned out by the shriek of jet engines, as the first squadron of Meteor
jet fighters started flying from there. As well as re-training pilots to fly
these new jet machines, there were also separate squadrons training
pilots to land on aircraft carriers; instrument flying; night flying; fighter
training; air attack and ground attack; anti-submarine attack, etc. One
division of the 52 Training Air Group alone consisted of almost 100
aircraft. With so many inexperienced pilots being fast-tracked through
the ranks, it is not surprising that there would be a few ditchings, but
these events do appear to have come in clusters.

1951 MAY The *W&S* was launched at 11.30pm on **May 15** and headed to a position 6 miles SW of Porthleven, where an aircraft was reported to have crashed. At 12.16 a message was received from Land's End Radio that the body of the pilot had been picked up. The boat arrived at the slipway and rehoused at 1.15am

1951 OCTOBER On a calm day, **October 4** with smooth seas, the *W&S* was launched at 1pm to search for a downed aircraft 5 miles South of Penberth Cove. After searching for about one hour, the boat was instructed to search about 1½ miles West by North of Tol-Pedn, and then 5 miles South by West of Runnelstone. Searched for about three hours, found nothing. Returned to station at 6.30pm.

1951 NOVEMBER The boat was launched again on **November 12** to search for another aircraft reported 2 miles ENE of Penzer Point. Searched for two hours and found nothing.

1952 JANUARY The first shout of the year came on **January 14, 1952** at 11.45am, after a Meteor jet fighter crashed at Mousehole and the pilot was seen to fall into the sea. The *W&S* was launched at 12pm, but as the lifeboat reached the spot a French fishing boat was rescuing the pilot, Lt Commander P.C.S. Bagley RN, who was taken to Penzance hospital suffering from shock and 'a few bumps'. Local press reported that the plane landed in a cow pasture, causing no immediate panic among its bovine occupants.

1952 JULY At 9pm on **July 18,** Penzance Coastguard requested the lifeboat to look for a rubber dinghy drifting offshore at Praa Sands with a young boy on board. The *W&S* arrived at approx 10pm and searched the area using a loud hailer and searchlight. At 11.15pm the boat received a message that the boy had been landed safe. The boat returned to station and was rehoused at 12.30am, July 19. The Cox'n noted 'J. B. Drew on holiday but also manned lifeboat'.

1952 OCTOBER A message was received at 20.41 on **October 8**, from Coastguards at Tol-Pedn, to be alert for an aircraft in the vicinity of Mousehole with engine trouble. The *W&S* launched at 21.25 and was requested to search an area between Mousehole and the Lizard. It searched until 22.45 when it was recalled to station by radio, arriving at Newlyn 00.20 on October 9.

On **October 25,** at 12.30pm the Penzance Superintendent of Trinity House phoned the Hon. Sec. to say that the wife of the master of the Sevenstones light vessel (*LV19*) had been taken seriously ill ashore and asked that the lifeboat would land him. As no other boat was available, the *W&S* launched at 1.10 pm in a rough sea, with a strong Westerly breeze. She took the master to Newlyn, arriving at 8pm. The weather was too bad for rehousing.

Maroons were fired at approx 9pm on **March 12, 1953**, following a message from the Coastguard that Chief Officer 19 Group, Plymouth, had reported that a plane was in trouble approx 7 miles 240 degrees from Loe Bar, and would probably fall in the sea. The W&S arrived on scene at 9.45pm and searched the area in company of the Lizard lifeboat, other HM ships and some merchant ships. Received several messages during the night from Land's End Radio to change position. Searched for 12 hours and returned to station at 9am, March 13. **1953 MARCH**

An urgent call from the Coastguard at Tol-Pedn on **March 27**, informed the station that an aircraft had dropped into the sea 2 miles SW of Porthleven, and that the pilot was in a rubber dinghy. The lb was launched, and about 15.45 received a message from Land's End RT that HMS *Eagle* had sent off a helicopter and had rescued the pilot.

The fishing boat *Boy Willie* of Penzance was reported overdue on **December 5**. Launched lb about 2pm, proceeded to the position in company with six trawlers, made a search. Poor visibility, about half mile. Fresh East wind and choppy sea. About 5.30 a coasting vessel named *Gem* located the *Boy Willie* about three miles North of Longships lighthouse. At 6 o'clock the trawler *Excellent* took the *Boy Willie* in tow. The W&S 'went and put our searchlights around and found everything OK.' After a wireless message from the trawler *Excellent* that all was well, the lifeboat returned to station. **1953 DECEMBER**

The 150th anniversary of the station on **March 12, 1954** was celebrated by a dinner held at the Queen's Hotel, Penzance. The commemorative vellum was presented by Lady Tedder to Coxswain Madron. During the afternoon the lifeboat was launched with Lord and Lady Tedder and Mrs Howard (wife of Cdr Greville Howard, MP) on board. **1954 MARCH**

On **March 13**. The French trawler *Aimable Madeleine* (L3897), leaving Newlyn harbour for the fishing grounds, collided with lifeboat W&S and local fishing vessels *Ocean Pride* (PZ134) and *Cynthia Yvonne* (PZ87) berthed at North Pier. Caused damage to all the foregoing, and proceeded to sea without reporting.

Open house at the Station

As the RNLI is funded by the generosity of the public, with no government or council support, relations with the general populace have always been very important. The main source of income comes from bequests, along with fund raising events throughout the country, including the many inland branches whose members might never actually see a lifeboat, much less witness one in action. Coastal communities have always been aware of the lifeboat and its value to themselves and to the international community of seafarers, while for many land lubbers a visit to the seaside on holiday or a day trip, is made complete with a visit to the local lifeboat station.

RNLI stations have always been encouraged to keep an open-door policy, with the public invited to look around the boathouse and, if lucky, view the boat itself. This would usually be restricted to looking in from a viewing platform, possibly walking around beneath the towering bulk of the hull or, on some special occasions, going aboard to see for themselves the working environment of the lifesavers.

The close up sight of these colourful machines, always immaculately polished and gleaming, has inspired generations of children who these days might come away with a souvenir model or item of merchandise from the station gift shop.

A party of happy children from Heamoor County primary school evidently enjoyed their visit to the Penlee boathouse in 1953 with headmaster Alfie Beckerleg

PHOTO: GEORGE HAWKES

During the 1950s especially, public interest in the lifeboat service was high and for those lucky enough to arrive at the time of lifeboat day, the chance to go afloat on the boat and take a trip around the bay or a run along the coast was the highlight of their holiday. The need to restrict access in more recent times, due to the sophistication of the vessels and stringent health and safety requirements, means there is less chance of taking a boat trip but, at the discretion of the Coxswain, lifeboats can be requested to take guests on special occasions, and be available for such events as the scattering of ashes, holding memorials and such.

The Penlee station's Lifeboat Day, Newlyn Fish Festival and the Lifeboat Open Air service have been annual occasions, and the boat has also been a regular attraction at the Porthleven sea festival.

Members of the public return from a trip out of Mousehole harbour (above and below) Left, a church service held on board at Newlyn with lesson read by St Ives MP, Greville Howard

1954 AUGUST The Hon. Sec. phoned on **August 2,** with reports of a yacht approx 6 miles ESE of Runnel Stone rocks with no sign of life on board. The Records of Service states: 'Lifeboat launched at 10.30am, and at approx 11.20 sighted a yacht sailing towards the Lizard. Informed Land's End Radio and was told by RMS *Scillonian* that another yacht was in position 2 miles SE of Runnel Stone Buoy. Altered course and went to the second position but found nothing. CG at St Just told us to give chase to a yacht sailing toward Lizard Head and see if he was alright, because a yacht is missing on passage from Penzance to Cork. At 1.15pm we told Land's End RT that we were slowly overhauling a yacht but it would take us several hours for us to catch him, and that it was being sailed perfect. At approx 1.45 Land's End RT ordered us to return to station.'

1954 SEPTEMBER There was a report on **September 3** that two men had been washed off the rocks at Rinsey Head. The boat was launched at 5pm. Arrived at Rinsey at 5.40. Searched for two hours, found nothing. The wind was moderate Southerly, with a heavy ground swell. Returned to station and housed the boat at 9pm.

1955 JANUARY A message was received at 11am on **January 4, 1955** that a ship was in a sinking condition off Porthleven. The *W&S* arrived at 12.15 to find the unnamed ship had entered Porthleven harbour. The lb returned to Newlyn harbour at 1.30pm. There was a moderate East wind with snow showers.

1955 AUGUST The Coastguard at St Just telephoned on **August 16** to say that a bather had been washed off rocks near Praa Sands. Maroons were fired and the lb launched at 12.10, arriving at Praa Sands at approx 12.45. Searched but found nothing. As the receiver on the wireless was out of order, the crew got in touch with people on the shore by loud hailer, and was informed by them that the bather had been rescued. The boat returned to station at 2.30.

1955 OCTOBER On the afternoon of **October 13**, 1955, the Port Medical Officer reported that a man on the SS *Manolito*, of Costa Rica, had been injured. At 4.15 the *W&S* was launched and went to Newlyn for a doctor. At five o'clock she embarked him. Then in a calm sea and thick fog she searched for the steamer. Arrived at position approx 6.15pm, stopped engines and listened for fog signal from the ship.

The lifeboat communicated with the Land's End radio station, using her radio telephone, and passed a message to the *Manolito* to blow a letter 'L' on her siren for distinction as there were other ships in the vicinity. Visibility was nil. The lifeboat felt her way to her and found

Manolito was built in 1943 as a tanker, named *Caribou* (US flag), converted to a dry cargo vessel of 7,210gt and renamed *Nathaniel B. Palmer* in 1946. After grounding in 1952 was repaired and renamed *Manolito* in 1953 (Costa Rica flag) owned by Marcou & Sons, London. Renamed *Manegina* (Lebanese flag) in 1960; scrapped in 1962.
PHOTO: WALTER E FROST

the steamer at 7.15pm, about six miles South of Penzance. She put the doctor on board the ship which was under way and steering SE by E. The lifeboat moored alongside the ship until 8pm, then re-embarked doctor and the patient and returned to Newlyn, arriving at 9.15pm.

The fog was too thick to allow the lifeboat to be rehoused, and she remained there until the next day. The owner made a donation to the funds of the Institution. Rewards to the crew, £14 5s; rewards to the helpers on shore, £9 18s. The *W&S* service record noted a 'small dent on the port bow'.

At 9.40 on the night of **March 1, 1956**, the St. Just Coastguard telephoned the Penlee station to say that the motor vessel *Crete Avon*, of London, a vessel of 4,000 tons, had been in tow of the tug *Cruiser*, of Glasgow, but that the tow rope had parted six miles East of Wolf Rock. At 10pm the *W&S* was launched. She made for the position in a heavy sea, with a West-by-South gale blowing and an ebbing tide. The *Crete Avon* was drifting eastwards and the tug was making for shelter to recover her hawser. The Porthleven coastguard Life-Saving Apparatus team stood by, and at 11.55 the Lizard lifeboat crew assembled.

At 11.30 the Penlee lifeboat reached the *Crete Avon*, which was then between nine and 10 miles West of the Lizard, and went alongside. The *Crete Avon* had on board 15 people, including a woman, but the master said that he needed no help and would wait for the tug. The lifeboat remained near her and passed a wireless message to the *Crete Avon* from Land's End radio station asking if she needed another tug. The master again said he needed no further help, but the lifeboat continued to stand by. The ebb tide eventually carried the vessel clear of The Lizard and to seaward, so the Penlee lifeboat returned to her station, arriving at 2.30 early on March 2nd. The Lizard lifeboat crew stood down at 2.25, and at 8.50 it was reported that the *Cruiser* had taken the *Crete Avon* in tow again.

The *Crete Avon* (3994gt) was built as *Heinz Horn* in 1928 at Elbing. In 1947 sold to Norway and renamed *Livarden*. In 1954 she moved to Crete Shipping as *Crete Avon* and in 1956 she became *Alderney*. Broken up December 1961.
JOHN B HILL COLLECTION

About 12.20 in the afternoon, the Lizard Coastguard telephoned the Lizard lifeboat station and said the tow rope had parted again and that the vessel was now three miles South-West of the Coastguard station. The master still said that he did not need a lifeboat, but at 12.57 the Coastguard reported that the vessel had drifted close in-shore and that the Life-Saving Apparatus team was assembling. At 1.20 the Lizard lifeboat *Duke of York* was launched in a rough sea with a moderate Westerly breeze blowing. She found the *Crete Avon* at anchor very close to rocks. The master asked the lifeboat to stand by, and she passed wireless messages for the vessel to the *Cruiser* and to Land's End radio station. The tug managed to take the *Crete Avon* in tow again at 3.20, and the lifeboat stood by until the tug had pulled the ship clear of the rocks. She then returned to her station, arriving at five o'clock. Rewards: Penlee, rewards to the crew, £16 3s. 6d.; rewards to the helpers on shore, £8 9s. 6d. The Lizard, rewards to the crew, £16 9s; rewards to the helpers on shore, £10 11s.

The *Warspite* salvage saga

What was purported to be the largest ever salvage operation in British waters lasted from 1947 until 1956. Following the drama of the *Warspite* rescue in 1947, the removal of 30,000 tons of scrap from Mount's Bay provided employment opportunities for the local community (and a tourist attraction for visitors) which lasted for years.

The vessel had been sold for scrap by the underwriters to a Mr Richard Bennett who had set up 'Western Salvage Co Ltd', of Penzance & Bristol. However, that company couldn't handle the job and it was taken over by The Wolverhampton Metal Company Ltd (WMC), who negotiated a contract in August 1947 with Mr Bennett, under which they were to advance monies against the vessel's non-ferrous metal to be salvaged and delivered, estimated at 1,250–1,500 tons.

The ship's bell and the partly scrapped hull (right) with tugs Englishman and Tradesman alongside

Early in 1949, the WMC took over the vessel and entered into agreement with contractors, 'P Bauer (Salvage) Ltd', to dismantle the remaining 25,000 tons of steel, non-ferrous metal etc. from the vessel, which was lying at Prussia Cove with a rock piercing her hull.

A cable car system, like a heavy-duty Breeches buoy, was set up to transport workers and equipment from the cliff top to the ship. A couple of clips from Pathé Pictorial and Movietone newsreels show this in action. In 1950 came the final attempt to refloat the hulk, using 24 Hollman compressors (supplied by Greenham plant hire) to pump air into the hull. Compartments within the hull were sealed and compressed air was pumped at a rate of 250 cubic feet per minute each — a combined total of 6000 cu ft per minute. (The supply of oxygen inspired the British Oxygen company to set up a depot in Cornwall). After 72 hours pumping the ship was refloated and towed to the beach at Marazion.

John MacQuarrie (left) goes to work by cable car
COURTESY MACSALVORS

Iain MacQuarrie (the son of salvage specialist Neil MacQuarrie) recalls that there were cranes on each side of the aft-mounted aircraft deck which had been used to recover Swordfish and Walrus planes that were launched from the ship by catapult.

In 1950 the Penlee lifeboat was called out twice to assist vessels working on the salvage. Two weeks after the first shout, on July 29, the salvage firm floated the ship and began to tow her away from Prussia Cove. However, the hawser fouled the tug's

The post at Prussia Cove was made from the ship's aerial mast and used as the base for the cable car

PHOTO: CHIS YACOUBIAN

propellers and the ship went aground again in an exposed position. On the advice of the salvage firm, all hopes of towing HMS *Warspite* to the breaker's yard on the Clyde were abandoned. Permission to beach her at a convenient site was refused and so demolition began where she was lying.

The WMC directly took over the salvage in January, 1952 when the contract with the salvage firm expired. The company had to start from scratch, building up a crew and acquiring the necessary equipment. Frank Wilson undertook this task with the help of Neil Macdonald and Neil MacQuarrie as salvage engineers and Duncan Nicholson in charge of plant. Meanwhile a search up and down the country was on-going for pumps, compressors, generators, cranes and steel air-lines and by April the WMC were contemplating their first major attempt to refloat HMS *Warspite*. While the firm previously employed had used four-tool compressors, WMC had the advantage of being able to use two jet compressors, loaned by the Ministry of Supply, and had expert advice from Rolls Royce Ltd.

On May 12th 1952, the WMC succeeded in lifting the ship and managed to move her inshore by a quarter of a mile. She then grounded and lay fore and aft across a deep channel and completely broke her back. In order to lighten the ship with the least possible delay it was decided to remove her 21-ton lower armour plates whole, and for this purpose a 25 ton crane was purchased. During this time the diver, Jim Craig, was repairing and sealing where necessary. Further attempts were made to refloat, but it was found impossible to lift the ship again so it was decided to sever HMS *Warspite* completely.

Another wreck became associated with the *Warspite* on December 23, 1952, when the Dutch steamer *Albatros* (1913/131dwt), carrying *Warspite* scrap from Penzance to Hull, went aground in thick mist to the West of St.Catherine's Point, Isle of Wight. The crew were rescued, but the ship was lost. There was also an episode in 1954 when a DUKW amphibious truck became stranded while taking out a party of surveyors who had to be rescued by a local boatman.

In May 1953 the forward end was coming away at the fracture. By June 27, she was finally severed and pulled inshore. During this period the WMC had recovered approx 5,000 tons of steel and 200 tons of non-ferrous metal. The plates were landed on deck, cut into quarters and sent by the company's barge to their Penzance Siding Depot to be railed to Sheffield.

As the operation was coming to a close the salvors, Messrs Macdonald and MacQuarrie, decided to consolidate their efforts and in 1957 formed the MacSalvors company which became the largest boat chandlers and heavy plant hire company in Cornwall.

Various artefacts from the ship are kept at Marazion Museum and other locations. MacSalvors still hold a section of the waterline belt at Pool, Redruth (Johnny Drew's 'torpedo blister' which the *W&S* had crunched into during the rescue).

A model of the battleship along with the ship's badge on loan to Marazion Museum, which features a display of Warspite memorabilia

The section of armoured steel waterline belt kept at Macsalvors is about 5ft square and 13 inches thick and weighs about five tons

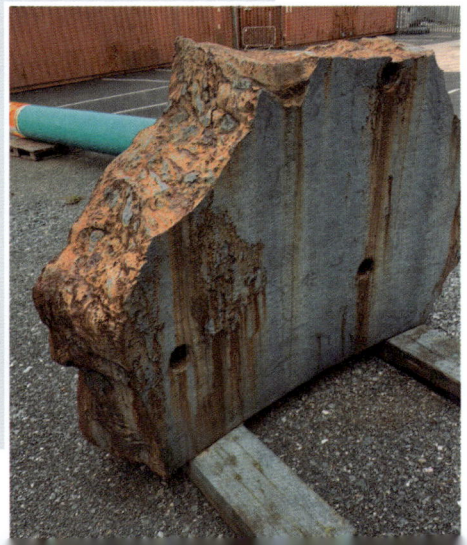

Seventeen crew lost at Wireless Point

By courtesy of   *Richards Bros.*

FRENCH TRAWLER AND PENLEE LIFE-BOAT

1956 MARCH At 5.45am on **March 14** the Coastguard received a telephone call from Porthcurno, to say a local man, out looking for driftwood on Porth Chapel beach, had come across a vessel on the rocks, with the number DE1517 on her bow. At 7am, the Tol-Pedn-Penwith Coastguard rang the lifeboat to say that a trawler had been wrecked at Wireless Point, Porthcurno. At 7.15 the *W&S* was launched. The sea was rough, there was a fresh South-Easterly wind, and it was one hour after high water. The lifeboat made for the position and found the trawler *Vert Prairial*, of Dieppe, on her beam ends among rocks. There was no sign of life in her, but several bodies were seen floating in the surf. The lifeboat picked up two of them and the others, which were too far inshore for the lifeboat, were picked up on the shore by LSA crew who were standing by. A helicopter arrived about 9am and several planes searched the area for another hour or so, but no survivors were seen. The lifeboat left for her station at 10.30, arriving at 11.15 and handed the bodies over to the police. All the trawler's crew of seventeen lost their lives. Rewards to the crew, £16 5s; rewards to the helpers on shore, £9 4s. 6d

The 250-ton French trawler *Vert Prairial* had left Brixham at 6pm on March 13. Nothing was heard from her again, other than a weak distress signal, prefixed by the call sign 'Dieppe 1517', picked up by the Scilly Isles steamer *Peninis*. That signal had also been received by the SS *Tiana*. It was assumed that she had gone ashore when her engine failed in the storm.

Captain Jean Baptiste Coppin and most of the 17 crew were found drowned, and their bodies were taken back to France in another French trawler. Newlyn locals remember that the local Catholic priest refused to bury the men together as some of them were not Roman Catholics, so a vessel from the trawler's owners was sent to collect the bodies. The dedicated team at H.N. Peake's Shipwrights & Undertakers, Newlyn, had the grim task of quickly making all the coffins. The coffins were placed on the deck of the French *Gay Florial*, sister ship to the *Vert Prairial*, and blessed by the Reverend Hosking from St Peter's Church, Newlyn, ready for a very sad repatriation to Dieppe. The coffin makers can be seen standing solemnly under the foredeck. Left to right: George Pomeroy, Nigel Brockman, Raymond Peake, Tom Waters and Nimrod Bawden. Nim later served one shout on the *W&S* and he and Nigel went on to become crew members together on the *Solomon Browne* lifeboat.

NIM BAWDEN COLLECTION

*Following the wreck of the *Vert Prairial* at the foot of the cliffs, it was shown that the wreck site was invisible from the Gwennap Head Coastguard lookout, which at that time was in the lower part of the present building. As a direct result of this wreck a second floor was added to give extra height for the watch room.

Soon after this service, the *W&S* was taken to Mashford's shipyard at Cremyll, where Johnny Drew spent almost six months completely refitting the boat (as described on pages 81-82). She was replaced at Penlee by the relief boat *Millie Walton* (ON848)

The relief lifeboat *Millie Walton* was launched at 4.15pm on **May 16,** on instructions from Coastguard St Just, after a request from CC Plymouth and CC Culdrose to search for a USAF B-47 Stratojet plane with three people on board that was down 6 miles SW of Wolf Rock lighthouse. The lifeboat arrived on scene at 6.45pm and searched in company with RN, RAF, USAF aircraft and ships from Plymouth. The sea was smooth with light winds. Searched until after dark. Left the scene at 10.15. returned to Newlyn at midnight. One survivor was picked up by a French fishing boat and landed at Penzance. One body was recovered later, while the search continued for the third person, who was never found.

1956 MAY

July 8: The relief lifeboat *Millie Walton* made a service to the SS *Yewcroft*, hard aground on Cudden Point. The 827 ton steamship stranded in dense fog on the rocks of Trevean Cove, while carrying cement between Cliffe (Gravesend) and Bristol. The captain believed he was near the Brisons at Cape Cornwall. Coastguard phoned the lifeboat at 7.10pm and the relief boat launched at 7.15 in thick fog. Received wireless message that the ship was ashore 3½ miles North of Cudden Point. The call was cancelled and a new message was that she was ashore 3½ cables N. While changing position the lb went aground on rocks known as Steven's Ledge. The visibility at the time was 'nil'. The lb was refloated with help from Mr Matthews in a small boat to run her anchor off and winch her off. No damage to lifeboat. Arrived at wreck at approx 8.30pm and was told by the master that a tug was on its way and asked us to stand by. At 10pm the ship suddenly broke in two and the crew left the ship. Took 10 survivors in the lifeboat and one other was rescued by LSA.

Certificate of Service and a Gratuity to Joseph J Madron, retiring after 1¼ years as Asst.Mech Penlee; 1½ yrs Mech Mumbles; 12½ years Mech Penlee; 9 yrs 2nd Cox Penlee; 9 years in crew Penlee.

The *W&S* was back on station, probably well before **November 6**, when a doctor phoned at 9.30am to say the 10,000-ton SS *Pontoporus* of Piraeus, then SW of the Scillies had on board a seriously injured man

who needed a doctor urgently. The doctor asked if the lifeboat would take him out. The *W&S* launched at 9.50am in slight sea and light E breeze and ebbing tide. At Penzance she embarked the doctor and ambulance staff at 10.20. Informed by Land's End Radio that the ship was in position 7 miles South West of Bishop Rock lighthouse and steaming at 16 knots. The Coxswain asked the *Pontoporus* to meet the lifeboat 5 miles S of the Wolf Rock, as it would take the lifeboat two hours to get to her position. At 12.20pm the doctor and ambulance men were put on board and the lifeboat accompanied the steamer to Penzance where the injured man was landed. The lifeboat left the ship at approx 12.30pm and arrived back on station at 2.20pm.

The *W&S* was launched in a South-West gale on **December 9** to take an injured officer off the 6,800-ton vessel *Harborough*. The officer was seriously injured after falling 30ft into a hold of the ship. Dr Leslie, of Penzance, who answers all calls from ships in the locality for medical assistance, was in the lifeboat, which launched at 12 noon. As the sea was rough, the *Harborough* was asked to go into

The *Harborough* was very new, recently delivered to Harrison Line. Her career ended in 1959 when she hit a wreck in the Weser estuary and sank

Mount's Bay, but even in the shelter of the land, it taxed to the limit the crews of both lifeboat and ship to get the injured man off. The injured man, named as Chief Officer Foster, was lowered in a sling by the ship's crane and landed gently in the lifeboat. He was landed at Newlyn at 3.15, along with a representative of the ship's master. The patient was taken to hospital where it was stated that his condition was serious. At 4.30 the lifeboat left Newlyn to take the master's representative back to the ship, and returned to Newlyn at 5.30. It was re-housed at Penlee the next morning

Medico made for television

A short information film, with a theme similar to the wartime productions, was *Medico*, made as a television play, which features the *W&S* in a short drama about taking a doctor out in the lifeboat to treat an injured seaman who has fallen into the hold of a cargo ship. (This might well have been inspired by the case mentioned above, the ship even has a similar name, *Besborough*.) The narrative showed the shore team in their control room, receiving the call for help from an approaching ship, notifying the medical team, and asking the Penlee crew to meet the ship two miles South of the Wolf Rock.

Scenes filmed on board the boat show the cramped conditions in the aft shelter, with the doctor and ambulance crew on board, prior to boarding the ship. The Medico service was dependent on ship-to-shore radio comms provided by the GPO, who funded this short film, made around 1956/7 and released in 1958. The cast of actors is listed with no mention of the lifeboat crew, although the *W&S* is seen in shot. The doctor has some convincing lines, however. "By the way, we'll need the Penlee lifeboat. Not enough light for the helicopter." And "I hate dodging about the Wolf in this weather. Makes me sick as a dog." It was thought this film had been lost forever, until it was unearthed in the RNLI archives by Elaine Trethowan, while researching material for this book. The producer was David E Rose, who went on to produce ground-breaking TV series such as *Z Cars* and *Softly Softly*, and later established the Film4 channel. Among the interesting period detail in this film is how much everyone, including the doctor, is smoking, either cigarettes or pipe.

Rehousing the boat with heavy TV camera mounted on deck aft

At 10am on **March 10, 1957** the St Just CG phoned that a rubber liferaft had been seen SW of Penzer Point. The *W&S* launched at 10.30am, found the liferaft was empty and brought it back to Newlyn. The boat could not be rehoused until two days later.

This is Your Lifeboatman

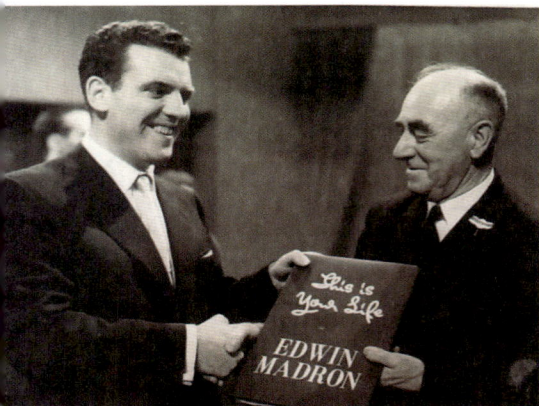

Back in the middle of the 20th century, there were only two television channels, BBC and ITV, broadcasting in black and white. To protect its dominance against this new competition the BBC introduced a programme called *This is Your Life*, based on a popular American show of the same name. The format featured a guest celebrity who was inveigled into appearing in front of a live audience, without prior knowledge. Presented by the genial Irish broadcaster Eamonn Andrews, this became a weekly high spot for many viewers who got a look into the private lives and background of the celebrities of the day: film and theatre stars, sports champions, war heroes, etc. The scripted dialogue was bound into a Big Red Book, which contained the subject's life story and would be presented to them at the end of the show.

1957 MARCH In those post-war years the lifeboat service had a very high public profile with annual flag days providing much of the RNLI's income, and the adventures of those heroes of the sea were frequently covered in the print media and cinema newsreels. Following the heroic efforts of our crew during the *Warspite* episode in 1947, the name of Penlee still resonated with the wider public ten years after it first grabbed the nation's interest.

On **March 18,** the spotlight was to turn on Eddie Madron, Coxswain of the Penlee boat and recipient of the RNLI's Silver Medal for Bravery. As a modest fisherman with no pretensions to celebrity lifestyle, Eddie was not the most typical 'victim', but he had an interesting story to be told — and retold. To synchronise with that year's Lifeboat Day, the programme researchers decided to focus on Eddie, and with the connivance of RNLI patrons Lord and Lady Tedder the set-up was planned. Lady Tedder had a personal association with Penlee and she had previously met Eddie Madron in 1954 when she presented him with the 150th Anniversary Vellum. The plot was to lure Eddie to London, to watch a Calcutta Cup rugby match on the Saturday before attending a lifeboat meeting on the Monday. As Lady Tedder told the press: "We told Eddie it would be a bit of a do about lifeboats. My husband might have to say a few words and Eddie might have to say a few words. And that's really what it was."

Eddie, who liked his rugby, was taken hook line and sinker (to reach for an appropriate fishing metaphor), and he was also an avid viewer of the TV programme and often wondered how they lured in their unsuspecting victims: He was about to find out.

On the stage walked Andrews holding the Big Red Book and announced to the theatre audience. "Ladies and Gentlemen. This is Your Life. Contained between the covers of this book is the drama and the sympathy which go to the making of a life. Tonight we tell the story of [he holds up book] Edwin Madron!

"We found our subject for tonight in a little community

300 miles from London. In one of the villages around our coasts among that breed of men without whom an island nation could never have survived, much less become great. The fishermen who so cheerfully go about their hazardous, daily business among our bays and harbours and out into the deep waters. [Film clips of coastline, calm seas and rough seas.]

""When gales beat against our shores, when fog or misadventure cause signals of distress the lifeboat crews of Britain launch their boats. And whenever the call may come, and however menacing the storm, at the first sound of the maroon, there will always be more volunteers running to the slipway than there is room for on the boat.

"Ladies and gentlemen, we have chosen one such man as the subject for our programme tonight, the eve of Lifeboat Day. From the gale-swept Cornish coast of Land's End, from the little fishing village of Mousehole, Edwin Madron, Coxswain of the Penlee lifeboat. There are 68 people alive today thanks to Edwin Madron and his crew and, although he has snatched so many lives from the cruel sea, that same sea has claimed his father and his son. In a moment or two he will enter the theatre, quite unprepared for what is to come — in the distinguished company of Lord and Lady Tedder, who have been in on our secret and have helped us getting our guest here this evening. Good evening Lord and Lady Tedder. It's a great honour to have you here in our theatre tonight. And it's a great honour also to welcome your friend Edwin Madron, because Edwin Madron: This is Your Life." Eamonn takes Eddie by the arm and leads him down stage.

Eddie in WWI naval uniform (below left). His mother Ada with brothers Ade, Eddie and Joe: the Renovelle (above), Lady Tedder boarding the W&S (left) in 1954 after presenting the 150th centenary vellum, and (below) Eddie meeting Capt Baxter, Master of the Warspite and his crew on the television show

ALL PHOTOS: EDWINA REYNOLDS COLLECTION

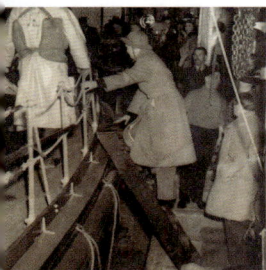

There is no surviving recording of the programme to judge Eddie's reaction, but it did not veer far from the script, as Eamonn took us through Eddie's early life, including the drowning of his father, who was Coxswain of the lifeboat; his early memories of playing at Mousehole with his old school friends and his time fishing on the *Renovelle*; his service in the Navy during World War 1, including time in the Falklands; his family dynasty and messages from his colleagues on the boat, most of whom had pre-recorded their recollections while remaining in Mousehole, standing by the lifeboat, ready as ever for the maroon to go off.

There were words from his brother Joe, and his daughter Stella was also in the studio, along with sons Joe and Jimmy, granddaughter Edwina, fishermen friends and Commander Wells in charge of HMS *Fort*, the base ship at Penzance. The Penlee treasurer, Owen Kernick retold the story of the *Warspite* rescue. Johnny Drew sent his recorded message, along with other members of the crew.

This led up to the high spot of the evening, meeting up again with the crew of the *Warspite*, in particular Captain Baxter who was in charge of the running crew. Eddie's wife held up his Silver medal to the cameras. Eamonn closed by saying "This is a tribute not only to you but to the thousands who man the lifeboats of Britain."

Master of the Warspite, Captain Alan Baxter.

1957 JULY

A message was received at 2.30 on the afternoon of **July 29**, that there was an injured man on board the Liberian cargo vessel *Cnosa* (10,000ton) bound for USA, about 20 miles South-East of Bishop's Rock Lighthouse. At 3pm the *W&S* was launched in a calm sea. She made first for Newlyn to take a doctor and an ambulance man on board and came up with the *Cnosa* 12 miles SSW of Penzance. The injured man was transferred to the lifeboat, which arrived back at her station at 7pm. The ship's owners made a donation to the Institution's funds. Rewards to the crew, £9 16s.; rewards to the helpers on shore £5 8s.

1957 AUGUST On **August 10,** the Coastguard called with a report of a ship sunk 3.5 miles E.N. of Runnel Stone Buoy. The *W&S* was launched at 6.30 but Cox'n Madron's log entry showed 'Nothing found'. The weather was bad, with a heavy sea and Wind Force 6 to gale. In communication with Land's End RT and a Shackleton aircraft also in the search, there was no sign of wreckage or oil. The lifeboat searched the area until dark and the search was abandoned about 9.15pm when Land's End Radio called the boat to return to station.

1957 SEPTEMBER The port doctor called the station at 6pm on **September 17,** asking if the lb could take an injured man off the SS *Alexandria*, which was then 150 miles from Bishops Rock lighthouse and expected to arrive 2 miles S of the Wolf Rock at 6am on the 18th. The *W&S* launched at 3am in smooth sea, fine weather. At Newlyn she picked up a doctor and ambulance man and made for rendezvous. At approx 6.45am she took the injured man on board and arrived back at Newlyn at 8.15pm to put the patient ashore. The service record shows 'slight damage to rubbing strake'.

At approximately 5pm on **September 24**, the station received a phone call from the Coastguard that a man was washed from rocks at Riddem into the sea. With permission of Hon. Sec., launched lifeboat at 5.15 in heavy sea. Arrived at 6pm, and searched the area with a helicopter. As the sea was too rough, returned to Newlyn at approx 7pm. The boat stayed there with a watchman for three nights and was re-housed on September 29. This was Eddie Madron's final service as Coxswain.

Eddie Madron (centre) hands over command of the Penlee boat to 'Jack' Worth on board the W&S. Edwin Madron retired after 10½ years as Coxswain; 12½ years as 2nd Coxswain of Penlee lifeboat. Awarded Coxswain's Certificate of Service and annuity. On the left is the new Hon. Sec. L.K.. Bennetts

L.K. BENNETTS COLLECTION

1957 DECEMBER At 10.45 on the night of **December 24,** the Hon. Sec. received a message that there was an injured man needing a doctor on board the tanker *Hemisinus*, of London, 150 miles South of Wolf Rock. A message was sent

MALCOLM CRANFIELD

Hemisinus *was a Shell tanker of 12,207gt, 19,575dwt, registered at London, Built that year (1957) at Cammell Laird. During the Vietnam war she earned the nickname 'Saigon Flyer' when trading Singapore to Saigon. She was broken up in 1976.*

back that the lifeboat would take out a doctor and meet the tanker 3 miles South of Wolf Rock lighthouse at 7.30 the next morning. At 5.30am on Christmas Day the *W&S* was launched in a smooth sea, under command of the new Coxswain, John (Jack) Worth, with the port medical officer on board. There was a light North-Westerly wind, and the tide was flooding. At 8.15am a message was received from the tanker that her position was 45 miles from Wolf Rock lighthouse. In reply it was suggested that she should make for Mount's Bay, where the lifeboat would meet her. The lifeboat reached Newlyn at 10.20am and then received a report that the 19-year-old patient, Roger Gibbings, was comatose and required a doctor immediately. The lifeboat left at once and met the *Hemisinus* 7 miles South-West of Penzance. The patient was lowered from the deck to the lifeboat and landed at Newlyn at 12.30, and the lifeboat returned to her station at 3.30pm. Rewards to the crew, £20; rewards to the six helpers on shore, £7 13s. The ship's owners made a donation to the Institution's funds.

On **July 17 1958**, there was a call from the Coastguard that a small fishing boat was in danger of breaking up on the rocks at Carn Dhu, Lamorna, and that a small boat had put out from Lamorna. Under the circumstances, Cox'n Worth thought it would be quicker to go out on his fishing boat *Porth Ennis* (PZ39) with three local fishermen, the brothers Pender. On reaching Penzer Point they met the fishing boat *Irene* with *Swiftsure* in tow. Took over the tow and brought *Swiftsure* in to Mousehole, with no damage recorded. **1958 JULY**

July 26 1958. Mr Barrie B Bennetts MBE died at the age of 76. Hon. Sec. of Penlee station branch 1913-1957. A solicitor, Registrar of the Penzance and Helston County Courts, a coroner for many years. He had played rugby football for England and represented Cornwall at cricket, golf and hockey. Mr Bennetts had recently resigned after 44 years as Hon. Sec. and was appointed an Honorary Life Governor, the highest award the RNLI can bestow on an honorary worker.

At **1958 AUGUST** 3.14 on the morning of **August 15**, the Coastguard told the Motor Mechanic that the motor fishing vessel *Hesperian* was ashore west of Lamorna. At 3.40am the *W&S* was launched in a moderate sea, with a gentle West-South-Westerly wind blowing and a flood tide. The lifeboat found the 20-ton fishing vessel in thick fog bumping heavily on the rocks close to Carn Dhu. A line was fired across to the *Hesperian*

EDWINA REYNOLDS COLLECTION

and a tow rope was passed to her. She was successfully pulled off the rocks and taken in tow to Newlyn, arriving at 6am. The service record notes 'BSA gun — 2 projectiles and line expended. Anchor lost and hawser cut. The Coxswain used his discretion and took one extra crew for this service' The skipper of the *Hesperian* expressed his thanks. The company which insured the vessel made a gift to the crew and helpers. Rewards to the crew, £9.12s; rewards to the helpers on shore, £5.8s.

This was Jack Worth's third service as Coxswain of the *W&S*. The skipper of the *Hesperian* was Joseph Brownfield, a lifelong Mousehole friend and fishing colleague of Jack's. They would have had a few laughs over this service! Joe and Jack are pictured together after the rescue.

1958 SEPTEMBER On **September 3**, the lifeboat was launched in response to a call from Coastguards that a person had been washed off the rocks at Praa Sands. Launched at 4pm and searched the area without success, along with two helicopters. At 5.25pm received a message to return to station. Arrived at Newlyn as the swell was too heavy to use the slipway.

A message was received from the Hon. Sec. at 12.45 on **September 16** saying that a fishing boat from Porthleven was overdue with three inexperienced men on board. Launched the *W&S* at 1.05pm, proceeding to Porthleven while keeping in radio contact with Land's End RT and Lizard lifeboat. At 1.45pm the lb was recalled as the fishing boat, *Renown*, had returned to harbour. There was a light SE wind, calm sea and fog patches. Arrived back at Newlyn at 2.30pm.

1959 FEBRUARY At 11 o'clock on the morning of **February 2, 1959** the port medical officer told the Hon. Sec. that the Cunard cargo vessel SS *Asia II* (8,723gt) of Liverpool, which was then three miles South-South-West of Carn Dhu, was making for Mount's Bay, as her bosun was seriously ill with pneumonia and needed medical attention. At 12.45pm the *W&S* was launched in a rough sea. There was a strong Easterly wind and it was high water.

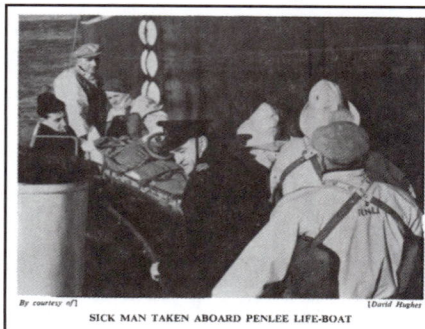

By courtesy of] [David Hughes
SICK MAN TAKEN ABOARD PENLEE LIFE-BOAT

The lifeboat made for Newlyn, where she embarked the port doctor and ambulance men, and left at one o'clock to meet the steamer. The doctor was put aboard the steamer, and after examining the patient he arranged for him to be transferred to the lifeboat. The lifeboat reached Newlyn at 3.15pm, when the bosun was taken to hospital. He was discharged 11 days later. The lifeboat remained at Newlyn because of the heavy ground swell near the Penlee slipway until February 9, when she returned to her station. The owners of the steamer made a donation to the Institution's funds. Rewards to the crew, £14; rewards to the helpers on shore, £10.19s.6d.

The *Asia II* was one of three 'A' Class cargo vessels built for Cunard's Canada service and delivered in 1947. It was the first post-war ship built for Cunard-White Star and their first cargo vessel. Its maiden voyage was to Canada in April 1947. In 1963 it was sold to the Eddie Steamship Co, Taipeh, and renamed *Shirley*. It was scrapped in 1968.

A phone message from Penzance Coastguard received at 8.30pm on **February 6** reported a light seen flashing 1½ miles SE of their position. Contacted Newlyn and Penzance pilots to see if there were any ships expected. They replied that none were due. The Hon. Sec. advised that the boat be launched to investigate the flashing light. Left Newlyn at 9.15pm and found a floating flare 2 miles South of Carn Dhu.

During the early hours of **April 1,** the Cox'n of the *W&S*, Jack Worth, **1959 APRIL** received a telephone call from the Hon. Sec. that a vessel was ashore at St Loy Bay, near Lamorna. The lifeboat was launched at 06.34. They proceeded towards the casualty, and when they were about half a mile away from the wreck, picked up a small boat with four survivors on board. The survivors indicated that a further two crew were still aboard the vessel, the French Crabber *Pluie de Rose*, owned by M. Kernionon. The *W&S* continued to the wreck and found that the remaining crew were being removed by Breeches Buoy. The Coastguard confirmed that all crew members had been accounted for. The *W&S* returned to Newlyn

Four of the crew clambered aboard the W&S and the lb crew recovered the ship's boat. Rare photos taken on board of crew members in action approaching a casualty. The shot on right shows Owen Ladner towards the bow, with Clifton Pender behind
PHOTOS: DAVID HUGHES

towing the small boat which can be seen in the photograph. They arrived at 07.45 and the survivors were taken to the Ship Institute (Fisherman's Mission). Rehoused lifeboat at 09.00. The crew were: Jack Worth, Coxswain; Jack Wallis, 2nd Coxswain; Johnny Drew, Mechanic; Clarry Williams, Assistant Mechanic; Clifton Pender, Signalman; Owen Ladner, Bowman; W Pender and G Beare, Crew.

Lloyd's List reported: 'French fishing vessel *Pluie de Rose* ashore at Trevedran Point, St Loy Bay. Tol-Pedn-Penwith and Treen LSA proceeding; Penlee lb called out. Four saved by lb, one swam ashore, one saved by Breeches Buoy. Skipper Albert Kernionon was the one who swam ashore and walked three miles to a farm to give the alarm. April 4 rapidly breaking up, No prospects of salvage.'

At 1.30am on **June 20** the Coastguard phoned with a message that a **1959 JUNE** white launch had left Newlyn for Porthleven, time unknown, but had not arrived. The lifeboat was launched at 1.42am and proceeded to the search area towards Porthleven without finding anything. At 4am it went into Porthleven harbour to find out more details. 'Was informed by the missing man's brother that he did not have any anchor, rope or matches on board. Requested CG to get in touch with CC Plymouth for aircraft to join search at daylight. Was informed that aircraft would be coming out at 0.800. Resumed search along with fishing boat *Chichester Lass*. He took a course along the shore and we went out to sea, thinking the launch had been carried away by the strong ebb tide. At 5am received a message from MFV *Chichester Lass* that launch *Mary M* had been found and towed into Porthleven.' The lifeboat returned to Newlyn at 6am.

1959 JULY The *W&S* landed a sick man from the SS *Lindi* of Antwerp on **July 19**. There was a moderate swell, and light SW breeze. The lb launched at 0.80 on the Sunday morning, embarked the doctor at Newlyn and met the steamer at Wolf Rock at 10.10. The sick man was transferred to the lb and they landed at Newlyn at 12.30pm.

Built in the USA and launched in 1946 as *El Salvador Victory*, the 10,000 ton ship was renamed *Lindi* from 1947-66 and traded for many years to the Belgian Congo.

The station mechanic received a phone call at approx 11.55 on **July 23** from CG Tol-Pedn with reports of a plane down 1 mile West of Cudden Point. Launched the lb at 12.07 and searched the area with helicopter and aircraft but found nothing. There was a WSW wind, Force 6 with rough sea. The aircraft gave up the search at 01.30. Lb continued on its own until 03.20 and then returned to station.

1959 SEPTEMBER

The *San Blas* was a reefer of 5,885gt/ 5,625dwt, built in 1955 for the Atlantic fruit trade, renamed in 1964 to *Olyutorka*. Broken up in Russia in 1977

On **September 12**, the lifeboat attended the 5,000 ton cargo ship MV *San Blas* of Stockholm and landed a crew man ill with appendicitis. At 7.35am the *W&S* was launched in a calm sea, with light E airs. The doctor and ambulance crew were embarked at Newlyn and then they made for the position 3 miles S of Penzance where the transfer was made. Arrived back at Newlyn at 8.45.

At 9.10 on the evening of **September 23**, the Coxswain was informed that there was a sick seaman aboard the tanker *London Resolution* of London. The tanker was expected to reach a position 5 miles South of

The 25,000 ton London Resolution (built 1958) was one of several London Overseas Freighter ships the young deck officer Rod Shaw served on. He was third officer on this ship in 1970 — a working lifetime away from his retirement project of restoring the W&S

Penzance at 1am. The *W&S* was launched at 11.15pm and picked up a doctor and ambulance crew at Newlyn. There was a moderate swell, a light West-South-Westerly wind was blowing, and it was two hours after high water. The lifeboat reached the tanker at 12.45am and the seaman, who was suffering from suspected coronary thrombosis, was transferred to the lifeboat. He was landed at Newlyn, where he was taken to hospital, and the lifeboat returned to her station, arriving at 2.30. Rewards to the crew, £9 16s; rewards to the helpers on shore, £5 8s.

The Hon. Sec. was informed at 3.25pm on **November 18** that the fishing boat *Girl Evelene* with one person on board had not returned to Newlyn. The weather by that time had made very bad, with wind SE Force 7-8. The *Girl Evelene* had not been seen by the CG at Tol-Pedn since 12.15pm, when she was trying to make Newlyn. The *W&S* was launched at 3.45pm and went as far as St Loy Bay, when Land's End Radio informed us that the *Girl Evelene* had made Sennen Cove and was being fuelled up on the slip. The lb returned to Newlyn at 5pm and moored for the night with the weather too bad to rehouse.

On the afternoon of **February 18, 1960**, at 12.48, the Motor Mechanic told the Hon. Sec. that a fishing boat had broken down close to the rocks at Tol-Pedn. At one o'clock the *W&S* was launched in a moderate to fresh SSW wind with a rough sea. It was an hour and a half before low water. The lifeboat found the fishing boat *May* (SS47) of St. Ives, with one man aboard, anchored 10 yards from the cliff face off Hella Point, Porthgwarra with a rope around her propeller. She took her in tow, but the rope parted several times before the fishing boat was brought into Newlyn. The boat reached her station at 2.15pm. Property salvage case.

At 1.10am on the morning of **May 8**, the port medical officer told the Honorary Secretary that a member of the crew of the Sevenstones light vessel (*LV19*) was ill and asked if the lifeboat could land him. At two o'clock, one hour before high water, the *W&S* was launched with the doctor and ambulance men on board. The weather was fine with a light South-Easterly breeze and a slight sea. The lifeboat reached the light vessel at 5.10. The sick man, H Semmens of Penzance, was transferred to the lifeboat and landed at Newlyn at 9.30am. Rewards to the crew, £15 17s; rewards to the helpers on shore £1 4s.

*During this service, *Lloyd's List* reported that emergency repairs had to be made at sea to the lifeboat's propeller. 'When seven miles from the lightvessel, the lb struck a piece of driftwood and her starboard prop was put out of action. Mechanic J. Drew lifted the special hatch over the prop and found the blades so badly bent that they were fouling the bracket. When alongside the lightvessel, while the sick man was being attended to, they straightened the blades with a hammer and crowbar sufficiently to enable the lifeboat to use this shaft at reduced speed on the homeward journey.'

A message was received at 10.15 on the morning of **May 19**, that there was a sick German seaman on board the anti-submarine frigate HMS *Undine,* and that a request had been made for the lifeboat to land him. The sick man had been taken off the Hapag-Lloyd cargo vessel *Saarland* 200 miles from Penzance. A rendezvous was arranged and at 8.15pm the *W&S* left Newlyn with a doctor and ambulance men. It was low water, and there was a moderate Easterly wind and a slight sea. The frigate was met 4½ miles SE of Penzance, where the seaman, who was suffering from appendicitis, was transferred to the lifeboat. He was landed at Newlyn at 9.25. The owners of the *Saarland* made a gift to the Institution's fund. Rewards to the crew, £8. 8s; reward to helper on shore, 12s.

This clears up the erroneous reports that the sick man was transferred to a 'submarine'. It was an 'anti-submarine frigate'. The only HM sub of that name was scuttled in 1940. The frigate HMS *Undine* was launched in 1943 as a U-class destroyer, and served in WW2. From 1952-1954 she was converted into a fast anti-submarine frigate, decommissioned in 1960 and scrapped in 1965

1960 MAY The Honorary Secretary was informed at 7pm on **May 25**, that a pregnant woman who was a passenger on board the Elder Dempster liner MV *Sangara* bound for Liverpool, from Lagos, Nigeria, had suffered a haemorrhage and needed a doctor. Following a radio call for medical assistance, the Penzance port health officer Dr Denis Leslie accompanied by two other doctors and his wife, an experienced nurse, and ambulance crew boarded the *W&S,* which left Newlyn at 8.10, half an hour after high

water. There was a light Southerly wind and the night was pitch black. The lifeboat met the *Sangara* 10 miles South-West of Penzance, where the doctors and medical team boarded the ship with difficulty, as it was now sitting in a swell that raised and lowered the lifeboat by about five feet. A blood bank was hoisted on board and a transfusion administered while the lifeboat stood off until called back by a blast of the ship's whistle.

After receiving medical attention the woman, Mrs Rita Akponwei, and her 10-month-old baby were transferred to the lifeboat, the child being lowered down inside a travel bag and the mother strapped to a stretcher. They were landed at 11.35pm at Newlyn, where an ambulance was waiting to take them to hospital. There the mother and child were reported to be satisfactory, although the unborn child was lost. The lifeboat reached her moorings at 12.15am. Rewards to the crew, £8 8s; reward to the helper on shore, 15s. The ship's owners made a donation to the Institution's funds. And, as Nim Bawden recalled, the crew received a welcome gift of four bottles of fine Guyana rum.

This was possibly the penultimate voyage for the *Sangara* (5,445gt) which was built in 1939 by Scott's of Greenock, for Elder Dempster Lines. Torpedoed by *U-69* on May 1, 1941, while lying at anchor at Accra, Ghana, and declared a total loss. One crew member was lost. The master, Sidney Themans, died while examining the wreck. Three months later an Italian submarine fired a torpedo at her but missed. On April 1, 1943 the wreck was sold for £500, refloated and towed to Lagos, and then to Douala, Cameroon, where the cargo was salvaged and sold. Elder Dempster re-purchased the wreck and towed it back to Lagos. The engines were overhauled, fittings renewed and the torpedo damage repaired. In 1946 the ship was towed to South Shields by the tug *Seaman*. At a speed of 2.5 knots they were underway for 62 days. After refit, the ship returned to the West Africa service in 1947. In September, 1960, four months after the Penlee medevac, she was sold to be broken up at Preston. The full story can be found at https://uboat.net/allies/merchants/ship/957.htm

Last shout at Penlee

On **June 15,** at 7pm, the *W&S* was called to land a sick person from the MV *Rowallan Castle* of London. She was expected to arrive in Mount's Bay at midnight. At 11.30pm the *W&S* put out with a doctor on board in moderate sea and fresh SW wind. She met the *Rowallan Castle* 4 miles South of Penzance and put the doctor on board. The casualty had a swollen left leg and after treatment was transferred to the lifeboat, arriving at Newlyn at 1.30am.

The *Rowallan Castle* was one of six small refrigerated cargo vessels in the Union Castle fleet, exporting manufactured goods to South Africa and returning to Europe with fruit. All five sister ships were of between 7,000-8,000grt, and built by Harland and Wolff at Belfast. *Rowallan Castle* was delivered in 1943 and consigned to the

N.W. EDWARDS COLLECTION

At the end of the month, on **June 30**, after 29 years solid service to the local community and to untold numbers of mariners, the old faithful *W&S* was placed on the RNLI Reserve list to make way for the 47-ft Watson *Solomon Browne,* the latest word in lifeboat technology.

The new boat was delivered from the builders in Littlehampton with a crew led by Cox'n Jack Worth and Mechanic Johnny Drew, who must have dropped a tear or two on leaving his companion of 29 years, while at the same time revelling in the sophistication of his new charge.

1960 JULY There were mixed emotions on **July 4, 1960** as the *W&S* left Penlee for the final time, making a farewell pass by the new boat, which was ready to take over its duties, and its place in the Penlee boathouse. Surprisingly, among such a sentimental community, there is little photographic evidence of that moment. But, as Johnny Drew might have said: "You can always repeat the truth. It happened".

The W&S steamed away from Penlee, bound for Scotland, and the doors of the boathouse closed on 'our' boat for the last time. As Penlee welcomed its new boat, the W&S would be open to new adventures in Northern waters

PHOTO: WATSON TREVASKIS

What had happened over those past three decades was that the boat had responded to 103 incidents, saving the lives of 102 mariners (with another 57 landed). She had brought home the bodies of more than 20 seafarers and spent many hours searching desperately for others who were never found. The boat had featured in two (possibly three) films and on prime-time television, and had participated in one of the RNLI's most celebrated services. Some 60 lifeboat crew members had selflessly gone afloat with full confidence in the boat, and there had been two medal winning services.

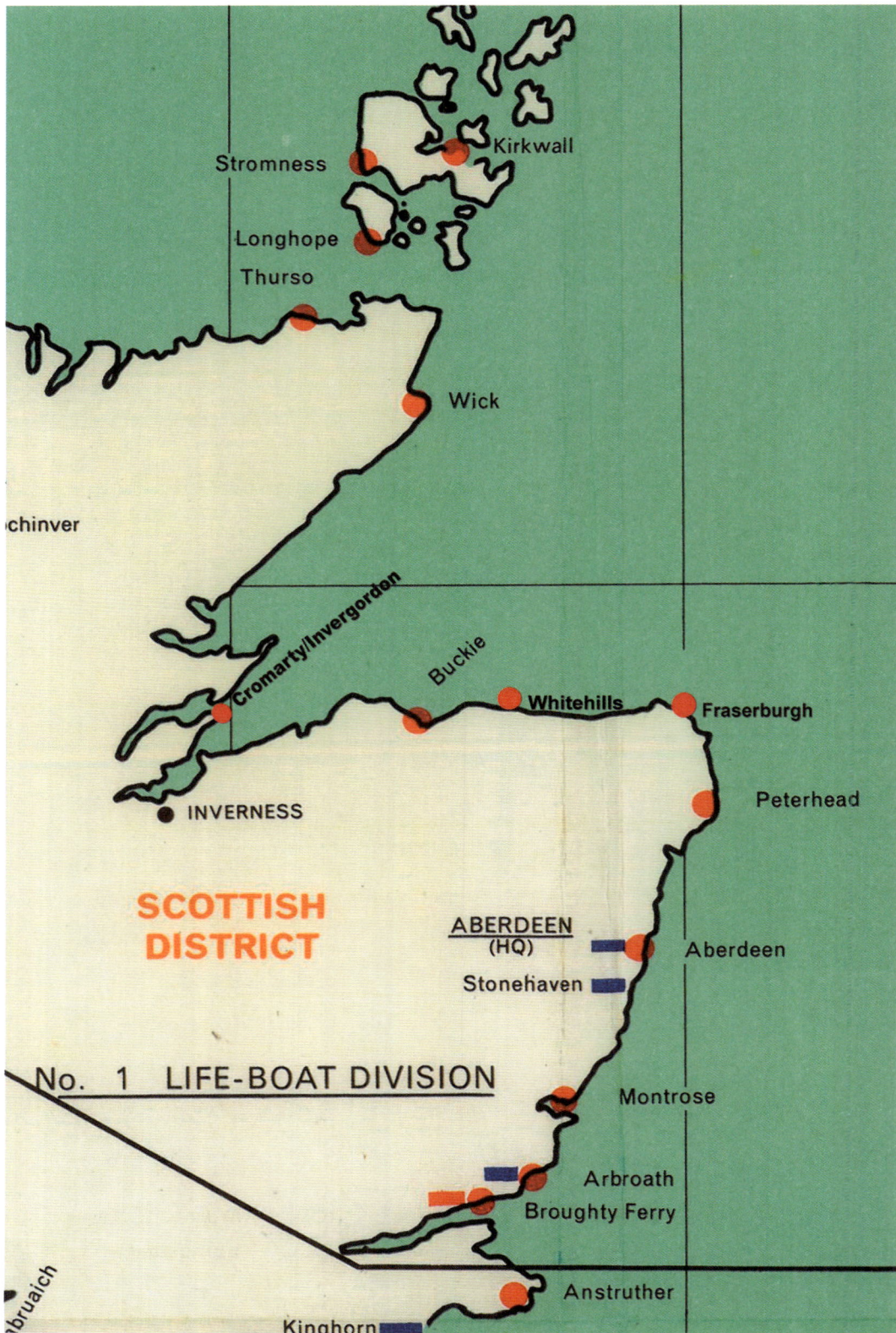

Stromness

Kirkwall

Longhope

Thurso

Wick

...chinver

Cromarty/Invergorden

Buckie

Whitehills

Fraserburgh

INVERNESS

Peterhead

**SCOTTISH
DISTRICT**

ABERDEEN
(HQ)

Aberdeen

Stonehaven

No. 1 LIFE-BOAT DIVISION

Montrose

Arbroath

Broughty Ferry

...bruaich

Anstruther

Kinghorn

Away to Northern Latitudes
via Land's End and John o'Groat's

1960 JULY

AFTER almost three decades in service at Penlee the *W&S* was to spend the next 10 years 'winding down' from her busy full-time career into semi-retirement in the north of Scotland, where she made only five services of any significance. On June 30, 1960, she was placed in the RNLI's Reserve fleet, to make way for the *Solomon Browne*. She left Cornwall for the North East Scotland port of Buckie on **July 4, 1960**, under the command of the Scottish Cox'n Tom Beattie, with Richard Richards (aka 'Double Dick', father of Trevelyan Richards, Coxswain of the ill-fated *Solomon Browne*) also on board.

RNLI records show there was often a full year between shouts, and during four years at one station, Thurso, she was called out only once. After 'departure from Penlee on June 30', 1960 she is listed as 'withdrawn' [being at Storeyard, Limehouse] 'Survey Relief Fleet from October'. However, we know that she was steamed directly to Scotland from Penlee, with Tom Beattie and Richard Richards, who brought home some Scottish souvenirs for his daughter.

Our boat's first deployment in Scotland was a temporary placement, in anticipation of the arrival of Buckie's brand new 47-foot Watson, *Laura Moncur* (sister of the *Solomon Browne*), due for delivery in June 1961. These latest versions of the well-tried and tested Watson design were taking over from the older 45ft 6 models in stations around Britain. As a relief boat, *W&S* was sent to replace the long-serving Buckie Lifeboat *Glencoe Glasgow* (ON857), which was then transferred to the Reserve Fleet in June and temporarily replaced by the Watson class *H.C.J.* (ON708), which had previously been on service at Thurso. A few days later she in turn was replaced by the *W&S*.

Buckie is situated on the exposed north-facing shore of the Moray Firth about midway between Fraserburgh and Inverness. It is a busy fishing port and home to several shipyards that built fishing boats and repaired all types of commercial vessels, including lifeboats, some of which were kept there while in the Reserve fleet awaiting assignment. The flank stations at the time were Whitehills (1924-1969) to the East and Cromarty (1916-1968, now Invergordon) to the West.

The station was opened in 1885. To accommodate Buckie's first motor lifeboat, a new lifeboat station had been erected in 1922, adjacent to the timber pier that marks the entrance to the harbour. Due to the restricted space, the lifeboat was launched from a cradle which raised and lowered the boat on a system of chains. The station was built of metal sheeting supported on a reinforced concrete frame and it remained in service until 1961, during which time considerable problems were caused by silting of the harbour, which often made

recovery of the lifeboat impossible at low tide. A new building was opened in time for the arrival of the *Laura Moncur*. As it was intended that the new boat would lie afloat at her mooring, a new gear store was built over a supporting platform of timber supports set into the harbour bed. The lifeboat was reached via an external stairway leading down to the floating pontoon. These facilities remained in use with little or no change until 1993.

The station records show that *W&S* launched three times on service during her first period at Buckie. There had been a service on **August 31**, a second shout on **October 1**, and another on **November 24**. The station comments that all shouts were 'routine', the most notable of which was the service carried out on October 1.

1960 OCTOBER At 9.32 on the night of **October 1, 1960**, the Coastguard informed the Honorary Secretary that red flares had been seen five miles North-West of Buckie. There was a light East-South-Easterly wind with a corresponding sea. The tide was flooding. At 9.55pm the lifeboat *W&S* made for the

position given. The light of the vessel which had put up a flare was seen, and as the lifeboat closed her, she was found to be the Sea School training vessel *Radium* of Aberdeen, which had developed engine trouble and was drifting west. The lifeboat, under the command of Coxswain William Stewart, took the *Radium* in tow to Buckie and then returned to her station, arriving at 11.40pm.

The converted Zulu type drifter was found drifting in choppy seas when the lifeboat arrived. A tow was established and the vessel was taken back to Buckie, arriving just before midnight. At the end of its life as a fishing boat, *Radium* had been bought by Robert Gordon's College in Aberdeen and turned into 'a kind of motor yacht' used for some years for teaching pupils seamanship and navigation. Round about 1960 her name was changed to *Radium of Don*. Originally home-ported at Findochty, the Banff registered vessel carried the number BF1489.

1961 JUNE

The Buckie crew on their new boat, left to right, Bowman John Innes, Alex Slater, Coxswain William Stewart, James Roy, Mechanic Jack Cole, James Murray and John Murray

Buckie station's archives record that: in **June 1961** *Laura Moncur* (ON 958) a new 47ft Watson class non-SR lifeboat, which cost £35,000, was placed on service, and the *W&S* was returned to the Reserve Fleet. The Buckie crew are shown aboard the new boat. (Interestingly, *Laura Moncur* was to finish her service life in the Westcountry. She visited Newlyn on passage in 1986 and again in 1987. She was then part of the Reserve fleet and her final posting was to Appledore, Devon in 1986-87. She was restored with the original external appearance as a leisure cruiser and was in good shape in 2020.)

The next posting for the *W&S* was to Aberdeen as relief boat, and it was over a year until she was next called into service. On **July 4, 1962**. The lifeboat was requested at 23.10 as red flares had been sighted about 3-4 miles ESE of Gregness. The lifeboat launched at 23.35 and was on scene at 00.40hrs with the Peterhead fishing vessel *May* (PD20), with a crew of three, which had fouled its propeller. The lifeboat towed the casualty back to Aberdeen, arriving at 02.30hrs.

The Aberdeen station was one of the earliest in Scotland, being established in 1802 by the Harbour Commissioners. A few days prior to this service, on June 30, 1962 Aberdeen No. 2 station and Torry Life Saving Apparatus had closed. The station's regular boat at that time was the Barnet *Ramsey-Dyce* (ON944). The Number 2 station had a carriage launched Beach/Surf ALB, the Liverpool class *George and Elizabeth Gow* (ON827).

The Aberdeen Cox'n in 1962 was Leo Clegg (right). He was a lecturer in sculpture at Grays School of Art in Aberdeen who had been awarded the DSC for his part in the famous commando raid Operation Chariot at St Nazaire during WW2. He was appointed second coxswain in 1960 and Coxswain the same year. He retired eight years later, after launching on service 40 times in all weather assisting 53 persons in distress.

From July 1963 until November, 1964 *W&S* was stationed at Cromarty, and yet another year had passed by before she was launched on **July 25, 1963** (R928) and no lives saved. The Coxswain of the Cromarty Lifeboat was Albert Watson and the Mechanic was his brother John Watson. (The Cromarty station, opened in 1911, was closed down in 1968, and later replaced by the nearby Invergordon station which opened in 1974.)

Our boat saw very little action during those years, but at Cromarty she was, at least, captured in photographs and, thanks to Ronald Young, Clem Watson, Arthur Bird and the Cromarty Archive website, we have tracked down what are probably the last pictures of her in RNLI service at the ferry berth in Cromarty Harbour.

The *W&S* then went to Thurso from November 1964, until August 31, 1968, with one launch recorded in four years. Thurso is the most northerly station on the UK mainland, covering the south Pentland Firth and north coast to Cape Wrath. Its exposed location means it is often

This postcard image is most probably the last published photograph of *W&S* in RNLI service, as discussed by knowledgeable lifeboat enthusiasts on the Cromarty Archive site. Tim Kirton wrote. "This is certainly a 45ft 6in Watson. I am almost certain she is marked 'Reserve Lifeboat' but it's not easy to tell from the pic. If she is marked Cromarty Lifeboat then she must be the *James McFee,* which was station boat at Cromarty from 1928 to 1955. If she is marked Reserve Lifeboat she isn't the *City of Bradford,* as that boat was unique in having a frame built over the windscreen arrangement in front of the cockpit; the frame was used to drape a canvas sheet over for added protection for the crew, fitted when she was still station boat at Humber. She isn't the *Julia Park Barry* of Glasgow or the *City of Edinburgh* as they were of the later 46ft 0in Watson design which had a modified 3-piece windscreen and funnel. Given the list provided by Quinton [Nelson] it therefore seems likely that she is the *W&S* which was a Reserve boat from 1960 until being sold out around 1968-69. She is now a yacht."

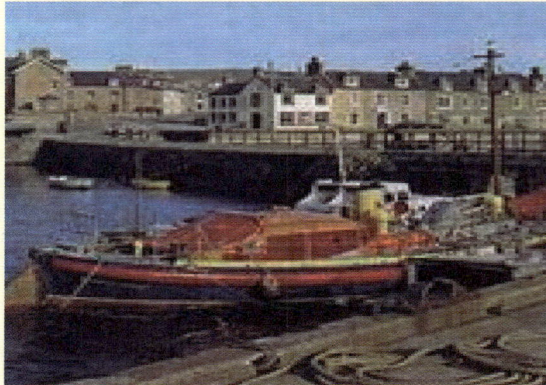

difficult to rehouse the boat. The flank stations are Wick to the East, Longhope (Orkney) to the North East, and Stornaway and Lochinvar to the West. These two are some way distant and situated beyond Cape Wrath. (That and Cape Cornwall are the only named Capes in the British Isles, both navigated by the *W&S* during her voyage north, which took her past both Land's End and John o'Groat's). The *Life-Boat* Journal described the coast of the Pentland Firth as 'steep-to cliffs characteristic of the area, which throw back the waves furious and confused, unmeliorated by beach or shelving sea bed. These can be tempestuous parts. The *Admiralty Pilot* speaks of winds off the north coast reaching Force 7 or more from 10 to 15 days a month in winter, increasing to Force 8 or more on about half those days.' Severe storms from the West are comparatively short lived, while gales from the South East can sometimes last for a week or more.

Thurso and its flank stations have as dramatic a history of Adventures and Perils as any in the RNLI's territories. The first lifeboat station was established at Thurso in 1860, when a 30ft self-righter the *Polly* arrived to serve the 'numberless vessels' that navigated the Pentland Firth. The first Cox'n John Brims served for 34 years, during which 304 people had been rescued by Thurso lifeboats. He was the first of 17 cox'ns to be awarded medals for their gallantry. As well as coastal trading ships, there were a great many fishing vessels that needed assistance — and there is still a very active fishing community to this day. A boathouse built in 1906 included a deep water slipway, just outside Scrabster harbour for the Watson boat *Sarah Austin.* In 1929 the first motor lifeboat arrived on station, the 45ft 6in Watson *H.C.J.* (sister to the *W&S*) and served until 1956. The replacement boat a 47ft Watson *Dunnet Head* (Civil Service No31) lasted only one year before it, and the boat house, were completely destroyed by fire.

At 6.45 p.m. on **January 12, 1966**, the Wick Coastguard informed the
Thurso Honorary Secretary that they had received a distress call from
the fishing vessel *Rowan Tree,* which stated that she was in a sinking
condition near Dunnet Head. The lifeboat *W&S,* on temporary duty at
the Thurso station, was
launched at 7.15 in a gentle
South Easterly breeze and
a slight swell. It was three
hours to high water. The
lifeboat went to the given
position and came up to
the *Rowan Tree.* A crew
was transferred to her.
Another fishing vessel, the
Leander, took the sinking
vessel in tow towards
Scrabster Harbour where
pumping arrangements had been made with the fishing boats *Primula*
and *Prospective.* The four pumps on board these two vessels were
transferred to the *Rowan Tree* when she went aground in the outer
harbour, and these emptied her of water sufficiently for her to refloat
herself. The lifeboat now took the *Rowan Tree* in tow to a safe berth near

the slipway, returning to her
station at 11.30 when the
fishing vessel was no longer
in danger.

FISHING BOAT
TRAWLER CAMPERDOWN OF ABERDEEN
LOBSTER BOAT ST. NINIAN
BODY

THE W. AND S. LIFE-BOAT

M.F.V. ROWAN TREE

The Coxswain at that
time was the highly decorated
Angus Macintosh DSM, MBE,
Croix de Guerre (right), who
served as Cox'n from 1937-
1939 and 1945-1967, and
received the RNLI's Thanks on Vellum for two services in 1953.

From August 31 until Sept 30 1968, the *W&S* is listed on the RNLI's
Lifeboat Integrated Computer System as serving at Holyhead (R884),
where the *Life-Boat* Journal reported she launched twice on **September
19** and **28**. The Holyhead station has no record of these shouts, but the
Cox'n at the time was Tommy Alcock and the Mechanic Don Forrest.

The *W&S* returned to Buckie for a second time from September 30, 1968
until April 30, 1969 with one more launch recorded on **April 20 1969.**

From April 30, 1969 until Sept 30, the *W&S* was stationed at Broughty
Ferry, Dundee, with three services recorded and six lives saved. The
flank stations here are Arbroath to the NNE and Anstruther to the SSW.
Ten years previously, in December 1959, the station had suffered one
of Scotland's worst lifeboat tragedies when their regular boat the *Mona*
(ON775), capsized with the loss of the complete crew. There had been a
furious storm raging across the British Isles, and along the east coast
of Scotland damage and devastation included power cuts, road and rail
closures, with ferries stopped and several ships storm bound. When the
North Carr lightship broke away from its anchors in St Andrews Bay
near Fife Ness at the mouth of the River Tay estuary in the early hours
of December 8, the *Mona* was launched to go to the rescue. Some time

between 5.15 and 6.00am the lifeboat capsized, her whole crew of eight losing their lives. The official report concluded that: "It is clear beyond doubt that the condition of the hull and machinery of the lifeboat at the time of launching were first class, and the engines and bilge pumps were working satisfactorily up to the moment of capsize. The crew were experienced and had complete confidence in the lifeboat and in her Coxswain, Ronald Grant." The upturned boat was found in one piece with crew still strapped in their seats but controversially the hull was set alight and burned on the beach, in keeping with a long-standing lifeboat tradition.

The *Mona* was an almost identical sister to the *W&S*; a 45' 6" x 12' 6" Watson cabin lifeboat with twin engines, each of 40hp. She was built by Messrs. Groves & Guttridge at Cowes in 1935. The report stated that 19 boats of this class were built between 1927 and 1935 [actually 23]. 'This is the first disaster to any of them. Crews have always spoken very highly of the sea-keeping qualities of these boats. The sister ship to the *Mona* based at Longhope, Orkney, crossed the Pentland Firth both ways against the tidal stream on the 7th December in a whole gale (Force 10

Crews have always spoken very highly of the sea-keeping qualities of these boats

to 11) which is strong evidence of the soundness of the design of the boat.' That boat, the *Thomas McCunn* (ON759), which is now preserved at Longhope, was coincidentally serving in the summer of 1969 as Relief boat at Penlee. (Tragically, the Longhope station suffered a similar disaster in December of that year, when its 47ft Watson *T.G.B.* (ON962) also overturned with the loss of all hands.)

1969 JULY At 12.15am on **July 2, 1969**, Dundee police informed the Hon.Sec. that a man had fallen from the Tay road bridge towards the Dundee end. The Inshore Rescue Boat (IRB) was launched at 12.19. It was half an hour before low tide. The *W&S* was launched at 12.25 to help in searching and provide parachute flare illumination. The IRB rescued the man from under the bridge and landed him at Marine Parade, where the police looked after him. The *W&S* returned to her station at 1.30am followed by the IRB at 2.15am.

The final call for 'our' boat

Our boat's last mission for the RNLI did have a touch of drama about it. A double shout, it involved rough seas, gale conditions, multiple casualty vessels, and six people landed safely. This day's events allowed the *W&S* to retire from the service with pride in its long-lived contribution to the saving of lives at sea.

The crew of the Broughty Ferry lifeboat in April 1969 with, in the foregound, Alick MacKay (white hat left) and John Jack (white hat right)

On **September 21** at 7.05am, there were reports that distress flares had been seen and it was learnt that the 38ft yacht *Sea Grim* was in difficulties near Lady Bank in Monifieth Bay, four miles East of the station. At 7.10 the IRB was launched in a strong gale force South Westerly wind, gusting up to 71mph, with a very rough sea. It was one and a half hours after low water. While on passage to the *Sea Grim* the IRB came up with the yacht *Fun and Games*, with three people on board, aground two miles East of the station. They were taken aboard the IRB and landed temporarily on a sandbank. A helicopter from RAF Leuchars, piloted by Flying Officer Paul Shaw, had also seen the flares while returning from another distress call, and he picked up the three men and landed them at Monifieth. The IRB then continued in search of the *Sea Grim* and found her out of control with eight people aboard. Five of her crew were taken aboard the IRB and landed on the shore, but the remaining three declined to be taken off the yacht as they hoped to make land safely. While the IRB was engaged in this service the trimaran *Nimble-Iki* fired distress signals off Tentsmuir but when the helicopter approached the boat the crew 'made two-fingered gestures' to indicate they did not require assistance.

The *W&S*, on temporary duty at the station, was launched at 7.51. She came up with the *Nimble-Iki*, which had by then got under way and did not require help. Following a report that the fishing vessel *Viking* was overdue, the lifeboat searched for her. She came up with that vessel with a crew of three, 200 yards East of the Horseshoe Buoy and took her in tow to Tayport. Five minutes later *W&S* was called out again to the assistance of the *Sea Grim*. She took the remaining three people from the yacht and brought them ashore. The IRB returned to her station at 8.55. At 12.45pm the *W&S* was back at her station.

The *Nimble-Iki* had got into difficulties some five years previously when being delivered to the Tay by lifeboat crewmen Laurie Anderson and Willie Findlay, and she was towed into Broughty Ferry by *The Robert*. The trimaran's owner, Ron Bonar was closely involved with the lifeboat station and eventually became branch chairman. To raise funds he had set up a beer festival which was named after him and raised around £200,000 over the 20 years following his death.

The Broughty Ferry Coxswain from 1964 to 1973 was Alick Mackay, who joined the RNLI at Anstruther in 1930 and was Mechanic of the Arbroath boat from 1957 to 1964. The Mechanic in 1969 was John Jack, who succeeded Mackay on his retirement in 1973 as Cox'n for the next 20 years. In 1986 Mackay and John Jack were awarded the British Empire Medal in recognition of their service.

*These were the *W&S*'s last services for the RNLI, bringing the total of lives saved to 108. From then until March 31, 1970 she was recorded [by Limehouse depot] as being 'Stored/Sale list' and 'Non-Operational'. She was then sold to John Marshal of Castlereagh, Northern Ireland for £1,450. And that's when her after-life began.

When the W&S steamed across the Irish Sea to Castlereagh, Northern Ireland in 1970 the boat was being prepared for a new role

A New Life at Leisure

FOLLOWING the final sequence of lifeboat services in Scotland during the 1960s, 'our' boat fell into a long period of decline and decay, before its partial renaissance another 16 years later.

On March 31, 1970 the *W&S* (ON736) was listed by the RNLI as 'sold out of service' to John Marshall, of Four Pines, Creamacreary, Castlereagh, Northern Ireland for a price that was the equivalent of £22,600 in 2019. He reportedly steamed her over from Scotland with his son Bertie, on behalf of the Sea Ramblers Sea Angling Club, based in Carrickfergus, but we have no record of the boat being used by them.

Having paid a substantial sum for this well-used, 40-year-old vessel, the owners apparently left it unattended for several years. The *W&S* could be expected to have been in good condition when the RNLI sold it on, but it seems the new owners never got round to restoring it, or keeping up with the essential maintenance and constant tlc that a wooden vessel demands.

Carrickfergus, County Antrim, is on the north shore of Belfast Lough, the intertidal water linking the capital city's port, at the mouth of the River Lagan, with the Irish Sea. The Lough is a long (21km), wide and deep expanse of water, virtually free of strong tides. As well as commercial shipping, there is a great deal of leisure boating in the area. The main coastal town on the southern shore is Bangor, County Down, where an inshore lifeboat is stationed, while the nearest ALB (All Weather Boat) is based at Donaghadee on the southern approaches to the Lough.

For a long time, the boat was tied up in Carrickfergus harbour by

The old W&S is barely recognisable in her new guise as the born-again Atlantic, *sailing out of Northern Ireland under new ownership in the 1980s, more than 16 years after her last shout as a lifeboat*

PHOTOS: SHIRLEY BODELL

two new owners, who it is believed were from County Tyrone, in the mid-west of Northern Ireland. The two men were ships' pilots in Belfast Lough, and they worked out of Carrickfergus pilot station on the boat, *Mabel Helen*. Coincidentally, a young couple who worked on restoring this boat after its eventual retirement were to become the eventual owners of the *W&S*.

The basic design and configuration of a Watson class boat does not lend itself to simple conversion into a personal leisure cruiser. The lack of access from the open helm position to below decks accommodation is a major disadvantage, and the separate watertight compartments compound that difficulty. Some similar vessels have been faithfully adapted in a way that retains the boat's original appearance but the most obvious change is to construct an enclosed wheelhouse closer to midships, directly above the engine compartment, and create an aft

"After one bad storm the boat broke loose of her moorings and sank

cabin in the space where the original helm position was. The obsolete and under-powered Weyburn engines would have needed replacing, along with gearboxes, prop shafts and other machinery. The simplest solution was to strip everything back to the bare hull and redesign the interior and topsides from scratch, which is what happened. At least the basic hull was sound and, as we will see, it was good to last for many more years to come. One of those enthusiasts working on the ex-pilot boat, Shirley Bodèll, was eventually to become deeply involved with the *W&S,* along with her partner, Basil Morton. As she recalled:

❝These guys basically left the boat there in the harbour, all plans to 'do her up' eventually abandoned because of the amount of travelling and time needed to complete their planned works. One subsequent owner, Tommy McKnight, tells me that the boat sat in Carrickfergus harbour on the East Pier for about five years, with fewer and fewer visits from the owners. And then no visits at all. Because she sat for so long, rubbing up and down the harbour wall, a hole was eventually rubbed into the port side of the hull.

"There was a boatyard, Carrickfergus Marine Construction, adjacent to the harbour at the time, run by another two pilots, Andy Brines and Jack Carroll. They could see that the boat was being badly damaged and, after one particularly bad storm, the boat broke loose of her moorings and sank. The local fire brigade were called to pump her out and re-float her, and the boatyard owners towed her up the ramp and along the harbour into their boatyard, where she sat for some months until they contacted the owners who lived out of town. The two yard owners weren't possessed of the skills to repair the double diagonal hull (mahogany on oak) but eventually they took on Billy Brennan, a shipwright from Harland & Wolff in Belfast, to work in the yard and he was capable of making the repair, which he did."

One of the yard's owners, Jack Carroll was an engineer, who fitted two replacement Ford D40 truck engines, which are still in place. However, sitting on the ground for all those years of neglect led to further deterioration to the hull. At some stage around then work began converting her into an offshore cruiser. This meant removing all the distinguishing features: the deck shelters, funnel, cowl ventilators, masts and fittings. The hull was painted pale blue and the boat was renamed *Early Mist*. When Jack Carroll died his widow sold the boat to Charlie Stitt of Islandmagee, County Antrim around 1983-4. Charlie added the wheelhouse and sold the boat on in 1986. **"**

In pale blue paint as Early Mist (far left) in the first stages of conversion, and (right) being repainted white by Basil Morton once he had taken ownership in 1986

Shirley's partner Basil Morton became the next owner of the part-converted vessel, and together the couple started a full interior renovation. They built a double master cabin, bunk athwartship, in the stern, two single berths in the forepeak, which were later changed to a shower cabin, WC and washbasin in anticipation of the owner's planned upcoming trips. A galley was installed aft of the shower cabin with a pot belly solid fuel heater (later changed to a Reflux diesel heater), table and fitted seating on the port side, and galley on the starboard side. The newly-added wheelhouse was fitted with seating on the starboard side and a big chart table. The ship's wheel was on the port side. Shirley recalls that Basil Morton never changed the design of the wheelhouse, which was his biggest regret: he hated that it was too straight/vertical. Each subsequent owner was of the same opinion but none has ever attempted to adapt or rebuild the structure, leaving the overall impression of being something of an 'ugly duckling', albeit with a degree of charm that oozes from its stylish hull shape.

Two water tanks were fibreglassed and fitted either side of the galley bilges to hold 120 gallons. The original fuel capacity was 50 gallons but more tanks were added in the engine room to give a capacity of 300 gallons. All the fuel tanks, existing and new, were steel. *On the hoist at Carrickfergus Marina*

There was a lazarette aft of the wheelhouse with access gained by the opening hatch on the deck. The couple fitted GPS, radar, autopilot and VHF. She was masted with main and mizzen and rigged, and given a good paint job, with white topcoat. Shirley later described the boat as "beautiful, finished and in very well maintained order". She was renamed *Atlantic,* with home port Belfast. Basil Morton owned the boat until 1997, and with Shirley he made good use of her, cruising local waters and even voyaging as far as the Mediterranean. As Shirley recalled when asked to contribute to this narrative:

A watercolour by Peter Bell of the converted boat, now known as Atlantic

"After Basil had bought the boat, he sailed her to East Belfast Yacht Club's slipway, in Sydenham, Belfast, where he was a member. He hoisted her up and started working on her. There was, however, a member of the club who demanded that Basil take the boat out of the club as she was deemed to be too long, as per the rules of the club, to remain on their hard standing. So, with more upheaval, we put her back into the water and sailed her to Bangor marina, County Down and stayed there for a few weeks and then, because of the expense of keeping her there, we took her to Newcastle harbour, County Down and remained there for a few weeks until the EBYC committee finally relented and let Basil take the boat back there. What a palaver! If not for this, the boat would have been fully refurbished and launched from there without all the upset and manoeuvres, from marina to harbour to boatyard. After the biggest part of the work had been done to the boat, Basil decided then to move the boat, and himself, into Carrickfergus Marina, in around 1989/90. He sold his home and moved onto her, where he lived happily until selling her on in 1997.

The boat was hoisted out every year and any necessary work was carried out, including antifouling, painting and replacing anodes. We did various trips on her, all commencing from Carrickfergus Marina: to the Isle of Man, Scotland, Wales, England, various harbours in the South of Ireland and of course our own Northern Ireland. We decided to journey farther afield and then, in 1995, we headed for the South of France, calling into harbours on the east coast of Southern Ireland en route: Arklow, Wicklow, Dunmore East, Rosslare.

On that voyage, the boat returned to Mousehole where it stayed for a week and was visited by several of the old lifeboat crew including the then Harbour Master, Frank Wallis, who had served 32 years with the Penlee lifeboat, and Douglas Blewett, son of Frank Blewett, the longest serving coxswain at the station. The *Western Morning News* reported that Frank Wallis was delighted at how well the converted lifeboat, then over 60 years old, was looking.

When we were sailing down towards Mount's Bay on the outward route, we were caught in bad weather, even though we had a good three-day forecast. It just came up out of nowhere. Very, very heavy rain and fog or mist to boot. And a pitch black night. We had no picture on our radar screen because the rain was so intense. This was 1995 and our radar wasn't the most modern then. We had GPS however — for a while.

So we're sailing on and the picture is not coming back on the radar and we're checking the charts when all of a sudden. . . a flashing red light.

Basil instinctively turned away from the light and I'm watching the GPS to correspond with the chart. Then, no GPS! We later learned (in Mousehole) that the GPS cable was caught around and severed by the steering gear when Basil was turning the wheel. So, no GPS and no radar. I was checking the flashing sequence of the light that was flashing bright red in front of us. It was the Tater Du light. The relevance of this is apparent in the case of the *Solomon Browne* lifeboat disaster. When it went to the rescue of the ship *Union Star* and was ultimately lost with all hands, the coaster was found west of the Tater Du lighthouse, capsized. And all hands lost too. We were in the very same vicinity. Basil turned 180 degrees away from the light and we just kept motoring until almost daylight. Eventually the heavy rain stopped and we saw a fishing boat heading towards us. We followed where he'd come from and entered into Newlyn harbour. Alive, tired and frightened. Another lifeboat disaster around the Tater Du was just too much to think about. But we were safe. At Newlyn the harbour master, Michael Haddock, came to collect fees from us. I told him we were the old *W&S,* come back to say hello. He phoned ahead to Mousehole and told them who we were. We motored on into Mousehole on the next tide to a rapturous welcome. They were waiting for us. It was wonderful.

We were so well looked after in Mousehole. Doug Blewett, the son of the *W&S* skipper, and his wife Ada; Cyril, ex-crew I think, the then harbour master Frank, most of the crew of the then lifeboat *Mabel Alice,* Neil Brockman being one of them, and so many more people, all with some kind of a tie to the *W&S,* they came in their droves. Neil took us to visit the old Penlee boathouse

and we were able to read of all the W&S's rescue missions and awards collected on the wall plaques and in the record books in the house. [Neil remembers going aboard and having a brief look around the boat with ex-crew member Frank Wallis. 'It was well looked after and well fitted-out, with a nice big wheelhouse'.] He, like so many other lifeboat

In Mousehole and Newlyn they were entertained by several of the old lifeboat crew, and were also observed by a much younger audience

191

crew of the *Mabel Alice*, had relations who died in the *Solomon Browne* disaster, and yet there they were, doing what their forebears had died doing.

We were wined and dined, driven to Penzance to do laundry, taken to Newlyn for gas refills, and anywhere we wanted to go really. We were taken by Doug Blewett on a drive around the beautiful coast, St Ives and all around that area and then to his home, where he sat us down and played all the video footage of the news coverage of the *Solomon Browne* disaster. Harrowing. We found it hard to pay for a drink in the pubs. We also found that no one, and I mean no one, would talk about the disaster to us, it was a very closed shop. Until they found out that we were the owners of *Atlantic*, the old *W&S* Penlee lifeboat. Then they, the locals, would talk forever.

We left eventually, new friends made, promises to keep in touch and off we went to France. We sailed from Mousehole and crossed the channel to Audierne. We worked our way down the Bay of Biscay, into Northern Spain, into La Coruna, Vigo, down the Portuguese coastline, calling into places like Viana do Castelo, Porto, Aveiro, Figueira da Foz, Sines and Peniche, Portimao, Faro lagoon and continuing eastwards through the straits and into Gibraltar. We stayed in Shephard's marina Gibraltar for about 10 days or so. Our first meal ashore was fish and chips!

We refuelled in Gibraltar and headed east along the Southern Spanish coast; Malaga, Almeria, Cartagena, Denia, Barcelona and on up the south east coast into the Golfe de Lyon into Agde. We were dismasted there and headed on into the Canal du Midi with the masts tied along each side deck. The guy who dropped the masts for us climbed up like a monkey in about five seconds, hooked all the shackles on and slipped down again. No crane, no lifeline! Dropped the masts in a short time. Away we went. Into the canal, first bridges. Very low! If we didn't get under this, we weren't going back home via the Canal du Midi!

It must have been around mid July when we entered the canal, as there were Bastille Day celebrations at that time in various towns. We exited the canal in Royan and got the mast stepped in Pauillac on the Gironde Estuary. We sailed north from Royan in Biscay and got home to Carrickfergus around August"

A few years later, in 1997, Basil Morton sold the boat to Tommy McKnight who kept ownership for a short while before he took it to Quinton Nelson from Donaghadee, who sold it on his behalf.

Very low airdraft on the Canal du Midi, but the boat managed to squeeze through with masts dropped

Return to home waters

IN 2000, some 30 years after retiring from service, the *Atlantic* was sold by lifeboat specialist Quinton Nelson, acting as broker on behalf of Tommy McKnight, to Gordon Burns, a businessman living at Falmouth, Cornwall. Quinton had brought the boat over to Bangor Marina from Carrickfergus for convenient access while painting and doing general restoration work, which included servicing the engines, cleaning fuel tanks, etc. After its previous white livery the boat had been painted emerald green all over. Shirley Bodèll remembered seeing it had been painted with, what she thought, was a bloody yard brush! "It broke our hearts. We couldn't look at her or even mention it to Tommy, as in 'What the hell have you done?'" Quinton painstakingly repainted it in RNLI colours where possible: the hull in Oxford blue, with red rubbing band and yellow gunnels, pale grey superstructure and red antifoul. She carried the RNLI logo transfer on the bow with Falmouth written on the stern.

Reflecting her lifeboat heritage, and now without the foremast, Atlantic had been restored by Quinton Nelson, who brought the boat back to Cornwall for its new owner

PHOTOS COURTESY QUINTON NELSON

Quinton Nelson had met Gordon Burns in June 1999 while he was staying overnight at Falmouth on his return passage from Poole, where he had attended the RNLI 175th anniversary event in his own Watson lifeboat. Quinton moored up in view of Mr Burns' house at Flushing, and when he came ashore the two got into conversation and Mr Burns expressed interest in finding an old lifeboat for his personal use. Quinton explained that an ex-lifeboat in original configuration was not really suitable for recreational use but if a converted boat became available that might be more practical.

When the *Atlantic* later came up for sale Gordon was interested, especially as it had local connections with Cornwall. He came over

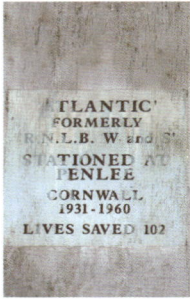

A faded sign on the cabin door gives a clue to the boat's history

to **Carrickfergus** and took it for sea trials. Before delivery, Quinton undertook the basic restoration including the paintwork and servicing. The boat was still fitted with a heavy, unoriginal mainmast and a mizzen, which did remain for a few years, before they were later dispensed with. Quinton recalled that he would have liked to replace the protective cork belting which had been removed long before, but the material was difficult to source and would have been hugely expensive.

Quinton delivered the boat to Falmouth via Milford Haven, Padstow and Newlyn, with his daughter Debbie and a friend, Ken Reid, as crew. As he recalled "This was during the foot and mouth epidemic and we were boarded off Wales, near the Smalls lighthouse, by police who climbed aboard from a RIB, unannounced and without asking, claiming to be looking for sheep!" When the boat arrived at Newlyn one afternoon, an elderly retired policeman, Derek Chapel, recognised it as the ex-*W&S* and came down to visit. He had himself been a crew member of the Sennen Cove lifeboat.

"Police climbed aboard unannounced claiming to be looking for sheep

At Falmouth, the boat was moored in the Penryn river on a fore-and-aft mooring, outside Mr Burns' house in Flushing looking over to the Falmouth Greenbank. Quinton later came across from Ireland to Cornwall a couple of times and accompanied Mr Burns in his own boat to the ex-lifeboat rallies at Fowey, but she didn't get much use as a cruising boat and functioned mainly as a floating office at Falmouth. As many boat owners discover, it was often difficult finding an extra person to help crew the boat, and coming on and off a fore-and-aft mooring is not for the single-handed. Just getting out to the mooring can become a pain in the (damp) backside.

Gordon Burns wintered the boat every year at a boatyard at Mylor Creek, off the River Fal, where Quinton visited on a couple of years to help with annual maintenance, including preparing the engines for the new season. But they always found it hard to paint the superstructure

which had been made of inferior plywood and absorbed a lot of moisture overwinter, meaning it took a long time to dry out.

From 2006 the boat remained ashore and it would not be returned to the water for several years. This was the second of the boat's extended periods lying ashore and it is a wonder that it survived at all. Boatyard owners generally despair of vessels fading away while taking up useful space, and *Atlantic* lay partly covered, at the back of the yard on reclaimed ground bordering some trees, where it gradually deteriorated, with rotting window frames allowing more rain penetration and damage. The water ingress seriously damaged the upper parts, decks and the interior, as far down as the engine room bilges, but it is thought the dampness in this particularly wet climate might have prevented the hull from drying out too much to be recoverable.

When the hull was stripped back by its next owner it proved to be in remarkably sound condition and, as Quinton insists "There was never any question about the quality of the hull; the older they are, the better materials were used. Fact! The conversions were done by handymen and later a joiner, not proper shipwrights who would have insisted on quality materials."

Both Mr Burns and Mr Nelson objected to news reports that the boat was found 'abandoned in a field', as they correctly claim it was being stored in a reputable boatyard. Nevertheless, seven years passed and the *Atlantic*'s structure continued to deteriorate and, finally, when the yard needed to free up the space, Mr Burns was asked to remove the boat. The *Atlantic* was eventually put up for sale and it was discovered in 2013 by Capt Rod Shaw, a lifeboat LOM from Harwich in Essex, who was looking for a lifeboat with an interesting history to restore. Rod recalls his first sight of the boat:

Looking forlorn and unloved in a Cornish boatyard, awaiting salvation

"As the inside of the W&S/Atlantic was still wet when I arrived, it clearly had been pumped out the night before. I decided this boat was worthy of a more detailed inspection and arranged to return some weeks later. I returned over a weekend and started a structural survey, as best I could, considering there was no electricity

or safe access to all parts of the boat. The adjacent boatyard did have good shipwright skills and everyone, from boatyard workers to boat owners, was very helpful and encouraging me to work on the boat there. This was quite tempting and I really enjoyed the Cornish friendship.

"Gordon Burns, the boat's owner, was asking in excess of £15k. Standing in the yard in pouring rain, with W&S looking beyond redemption, I offered him less than half of that and was prepared to walk away, as the costs for the salvage and transportation back to Harwich were considerable, even before restoration began. Finally, he relented and accepted my offer. The challenge commenced! I returned home to plan the campaign. **"**

Atlantic at the Mylor yard, And (below) finally leaving for pastures new

Rod's logbook, in the concluding chapter, brings the boat's life story up to the date of its 90th birthday.

*Quinton Nelson is acknowledged as *the* lifeboat expert, who has restored many of the disposed RNLI boats at his business in Donaghadee, N Ireland. He and his father before him were lifeboat men and he has an interesting personal connection with Penlee. Since 1988 he has owned and operated the classic Watson 46ft 9in boat *Guy and Clare Hunter*, which took up relief duties at the Penlee Station in 1982 after the loss of the *Solomon Browne*. Built in 1954 by J. Samuel White of Cowes (the makers of ON736), *Guy and Clare Hunter* (ON926) served at the St.Mary's Station on the Isles of Scilly from 1954 to 1980 and was involved in many high profile services including the *Torrey Canyon* disaster in 1967, 'standing by' for several days, and alternating with the *Solomon Browne*, while the salvage teams attempted to refloat the tanker. When she was replaced at Penlee in 1983 she went on to Padstow. She was finally sold on from the RNLI in 1988 as a pleasure cruiser. The *Guy and Clare Hunter* is the only 46ft 9in Watson lifeboat afloat in completely original service condition, and there are hopes that one day she might possibly return to the Penlee boathouse in glorious retirement.

Resurrection at Harwich

THE boat had experienced two periods of renaissance, but those were only semi-permanent, and if the *W&S* was to endure past its ninth decade it called for resurrection. This was its last chance to escape the breaker's yard and on August 19, 2013, the *W&S* departed

Cornwall bound 'on passage' for Harwich, by road, rather than water. Malcolm Elvy's specialised low loader hauled its oversize cargo on the 364-mile journey from the far west of the mainland to its new home in the far east, arriving at Gas House Creek on August 20.

The small town of Old Harwich is very much a maritime environment. It is the focal point of Harwich Haven, which includes the

UK's largest container port at Felixstowe, along with Harwich Navyard, Harwich International at Parkeston Quay, the river ports of Ipswich and Mistley and the Trinity House pier. To the west of the Trinity House buoy yard lies Gas House Creek, the site of an earlier boatyard and set to be the new home of *W&S*.

To the east of the Trinity House pier the town's Quay is marked by the smart, 21st century lifeboat house, built in 2003, which takes a prominent position, with its Severn class ALB *Albert Brown* highly visible afloat on its pontoon mooring, the ILB housed at street level, a public viewing gallery and souvenir shop. There is also a small independent lifeboat museum around the corner next to the pilot boat moorings, occupying the 1876 boathouse.

The Harwich station has, in some recent years, been one of the busiest in England, and it is one of the oldest, having been set up in

The W&S rolls into Harwich and is backed onto the Gas House Quay, where the young Rod Shaw once worked as the 'Dolly Boy', crawling beneath the hulls of vessels under construction or repair — a situation he was about to revisit

The Duke of Northumberland was stationed here from 1890-1892, credited with saving 33 lives

1821 by the Essex Lifeboat Association, although its first boat was withdrawn in 1825. It can claim several firsts, including the first steam-powered lifeboat, the *Duke of Northumberland*, and the first Severn class lifeboat, *Albert Brown* (17-03).

In 1876 the RNLI sent the *Springwell*, a self-righting type, 35ft pulling and sailing boat, crewed by 10 men. In January 1881 the boat launched to the aid of a Dutch vessel, saving seven lives from the wreck and the Coxswain and crew were awarded medals by the Dutch Government. Tragically, it turned turtle later in 1881 with the loss of one crew member. It was immediately replaced by another boat of the same name, which served at Harwich until 1902.

Following that was a 43ft Watson class lifeboat, the *Ann Fawcett*, which stayed at Harwich until the No.1 station was closed in 1912. In 1890 a second Harwich Lifeboat Station had been opened. Their first boat, the *Duke of Northumberland*, built in 1889, was the first ever steam-driven 50ft lifeboat and was a revolutionary water-jet design. During its short stay at Harwich it was moored afloat in the Pound near Halfpenny Pier. This boat was transferred to Holyhead in 1892.

Replacing the *Duke of Northumberland* was the *City of Glasgow*, built to the same design but 2ft longer, which was designed by the Institution's Naval Architect G.L. Watson. A second steamboat of the same name followed, but in 1907 the station was closed due to the Admiralty requisitioning the steam tug which assisted the lifeboat during services. The port then went without a lifeboat for almost 60 years, with local waters served by flank stations at Walton-on-the Naze and Aldeburgh. Trinity House pilot boats were based at Harwich and would have taken on much of the search and rescue role.

With the increase in recreational boats in the rivers Stour and Orwell, and following two separate drownings, the local pilots organised a public fund for the provision of a rescue boat at Harwich. During evaluation of small, inflatable boats the RNLI decided to trial an all-weather boat, which resulted in the re-opening of the Harwich station in 1965. The most notable incident in recent years occurred in December 1982, when the ro-ro car ferry *European Gateway*, leaving Felixstowe for Zeebrugge, was in collision with the train ferry *Speedlink Vanguard* approaching Harwich Harbour. The *Gateway* capsized and despite the successful efforts of pilot boats, ship-handling tugs and the Harwich lifeboat in recovering most of the casualties, six people were lost.

In the first decades of the 21st Century there were several years with more than 100 shouts clocked up by the station's Severn class ALB and Atlantic ILB. As part of the Harwich Haven group of commercial ports there has been a variety of shipping incidents. Ultra large box ships berth at the UK's busiest container port of Felixstowe, with feeder vessels, tugs, dredgers and ro-ro ferries competing for space. Trinity House has its buoy yard there (having moved out of Penzance some time earlier); there is a fuel tank farm at Parkestone; a Stena ferry terminal

The City of Glasgow was a Watson design, seen here at the Pound in Harwich

linking with Hook of Holland and a berth for cruise ships; short sea bulkers come down river from Ipswich; wind farm traffic has been based there and a fleet of fast pilot launches serves the busy Sunk pilot station, 16 miles offshore. On top of all that, there are an estimated 10,000 berths for leisure craft in the surrounding rivers and backwaters.

The new owner of our boat, Captain Rod Shaw has been actively involved with the Harwich Lifeboats since July 14, 1975.

The European Gateway / Speedlink Vanguard collision in 1982 was a major disaster which led to improvements in the safety of ferries, helped by Rod Shaw's participation, investigation and later large scale safety exercise

PHOTO: IPSWICH STAR

❝I remember this date because it was the day I left the Merchant Navy and started a job with the Harwich Harbour Master. I represented him on the local lifeboat committee and became DLA (Deputy Launching Authority). Because we had so few crew I often ended up doing a lifeboat shout myself and have earned a number of commendations. One was for a shout on the *Albert Brown* (17-03) lasting 22 hours when we ended up in Zeebrugge, Belgium, having experienced a force 11 storm. The shout took us 55 miles off Harwich but due to the weather and sea conditions we considered it better to make for Zeebrugge. Another was to a trawler on fire when the Harwich Lifeboat crew battled in strenuous firefighting efforts. As time progressed I became Hon. Sec. which is now called LOM (Lifeboat Operations Manager). Also, my wife has been secretary of the Ladies Lifeboat Guild and my son Brendon is by now a senior member of the Harwich crew.

As I approached 65 years old, with retirement looming, I realised I needed a hobby or pastime to occupy me. I had worked hard all my life and the only interest I had was lifeboats. By this time I had received the RNLI Gold Badge and Bar to Gold for services to the RNLI. I was delighted to be awarded MBE in the 2013 Queen's Birthday Honours for services to Maritime Safety. Receiving that honour coincided with the purchase of my own boat. My life had entered yet another phase.

Retirement really worried me, I did not want to be sitting around the boathouse each day waiting for the next shout. I kept deferring retirement. I had never been interested in owning a boat, even though I was a local fisherman before I went deep sea. It was my wife who suggested I got a boat to do some lobstering in my retirement. This led me to thinking about an ex-lifeboat which I could also display on lifeboat day. I needed a boat that demanded work that was within my practical capabilities. This became an exciting project for retirement and led me to search for a suitable boat.

The first Severn class boat in service, Albert Brown *17-03 on the Lifeboat house pontoon*

I did all my sums, because once you retire you do not have a continuous income, and a wooden boat more than 80 years old would demand some serious money. But my wife had realised that this project would be the answer to a happy retirement and she encouraged me to take the project on, provided it was achievable.**❞**

Captain Roderick 'Rod' Willis Shaw FNI, MBE

Rod is a Master Mariner, having spent a lifetime at sea or within the maritime industry. He was born, bred and still lives in Harwich. From working as a boy fisherman and lobsterman he gained a

four-year apprenticeship in the British Merchant Navy, serving in cargo ships, tankers and bulk carriers and often doing long voyages of up to 12 months. His early Merchant Navy career was spent on tramp ships trading world wide, including supplying fuel to the American forces during the Vietnam War, and he served as navigating officer on what was, at the time, the largest ship in the world, the VLCC *London Pride* (125,337gt/255,090 dwt). In 1970 Rod was serving as Second Mate on the same company's vessel *London Resolution*, which had been the subject of a medevac service from Penlee 11 years previously. His early years followed a pattern very similar to those of the Mousehole lifeboat crew.

Rod was appointed MBE in 2013, the year he purchased the W&S

"As a boy I mucked in with the local fisherman, of which in the 1960s there were quite a number fishing out of Harwich (shrimping comes to mind). In my teens I spent more time on boats than I should have done and my education suffered. I was by then earning money as a boat hand on the crew tender serving the laid up ships in the River Stour. This was further complemented by taking out sea fishing parties and lobstering during the summer months. My mother was adamant — in fact had very strong views — that I was not going to be a Harwich fisherman.

Having left school without any qualifications, career options were somewhat limited. Mother agreed that I could go into the Merchant Navy as a trainee officer. But without exams I was not accepted for pre-sea training at any naval establishment. In addition, no British company would consider me as I had no exams to my name. A local pilot who I met said his son was with London and Overseas Freighters and I should try applying to them as they were not as stringent as the UK liner companies. Even BP tankers turned me down!

I had been working quite a lot for the Harwich Harbour Master on his launches and he agreed to give me a reference. I treasured this piece of paper as it was the only positive asset I had. Anyway, I sent it to LOF who called me for interview and the only question they asked was 'Are you a practical person'. I could show that I was by my boating involvement and they offered me a four-year apprenticeship. (Mother was so happy!). My pay was £17-10s per month with an additional shilling (5p) in lieu of washing. I soon found out why they were looking for practical persons as I spent the next four years on deck and got all the jobs they would not give the Indian seamen. An example of this was on the cargo ship *London Statesman* when we were loading military supplies in America and the longshoremen (stevedores) would go to the toilet in the cargo hold and it was my job to shovel it up and take it out of the hold. The Indian crew refused to shovel shit, which is why it was left to the apprentices.

London Independence was another of Rod's ships, seen in dry dock at Falmouth in 1973

An apprentice on four-year indentures had to complete three years seatime. I completed my seatime in three years and three months because of my long voyages and also short leave periods between ships — by mutual agreement, as I was happy to cash my leave in and go back to sea. On completion of my seatime I was promoted to uncertificated 4th officer and suddenly found myself on the bridge, which was a new experience. I was on watch with the Chief Officer one day when he went off drinking with the captain, leaving me in charge. A warship passed and on the Morse light he signalled 'You are steaming into . . .

something?' I could not read Morse code, so I called the Chief Officer and explained about this warship which was steaming hell for leather in the opposite direction and was now over the horizon. Next morning we ran straight into a hurricane and the Captain called me out of my bunk and said "You got us into it, you can see it through". I stayed about 18 hours on the bridge as we went straight through the eye of the storm.

Another warship experience was on the supertanker *London Pride*. I had transferred from the oldest and smallest ship in the fleet, *London Harmony*, to this massive, 340 metre long tanker, joining in Le Havre. I received no extra training or preparation. I was in awe at the sheer size of this monster and I was in charge of the 8-12am watch. Anyway, just off Ushant this small French warship was crossing on my starboard bow on a steady bearing and it was my responsibility to alter course out of his way. I was so frightened to swing this large ship, eventually the warship came round at the last minute and kept clear.

One time we were in the Persian Gulf in reduced visibility due to heat haze. The captain was on the bridge, when out of the haze came another VLCC, Japanese flag. Our supertanker had a panic control to bring the ship hard over. The captain hit the control and we just missed the other ship. Had we collided it would have been one of the biggest collisions in maritime history; the combined tonnage would have been 550/600,000 tons and he was fully loaded with crude oil.

As a young man Rod was Navigating Officer on the VLCC London Pride, at the time the biggest ship in the world

A number of the older tankers were converted to bulk carriers, which traded as tramp ships around the world. On one of them we loaded 15,000 tons of wheat by dockers carrying sacks up the gangway, slitting the bag and pouring it into the hold. On discharging in South America, the dockers used biscuit tins to fill sacks which were sewn up in the cargo hold and then lifted ashore. It took over two weeks to discharge. **"**

Following a slump in shipping during the 1970s, Rod progressed his career in the UK ports industry as: Port controller in UK's busiest commercial harbour, Harwich Haven; Hydrographic surveyor; Marine Pilot; Harbour Master of Harwich Harbour (becoming the youngest Harbour Master of a UK major port). He became a specialist in port safety and navigation in congested waters, and represented the UK Ports Industry at international level. He was also adviser to Marconi and other radar manufacturers on port radar systems. He received a commendation for actions following the *European Gateway* and *Speedlink Vanguard* disaster in 1982, when two North Sea ferries collided outside Harwich Harbour and a major rescue was required. Subsequently, to learn lessons from that disaster, he organised the country's largest emergency exercise involving a ferry with 600 passengers.

His career progressed to become Marine Adviser to Lloyd's Register of Shipping, and representative of International Shipowners at IMO (International Maritime Organization, which is the United Nations department for maritime rules and regulations); Nautical Surveyor for MCA Department of Transport; Examiner of ships' captains and navigating officers and Inspector of foreign ships using UK ports; Surveyor of ships flying UK Red Ensign; Investigator of accidents and incidents on River Thames and Essex rivers.

Rod's MCA highlights include: Active participation in Queen's Diamond Jubilee River Pageant in 2012; Lead surveyor on Government's implementation of Boat Master Licences; Chairman of River Thames Local Knowledge exams; London and Southampton examiner of ships' captains and other deck officers; Senior MCA Nautical Surveyor on the River Thames.

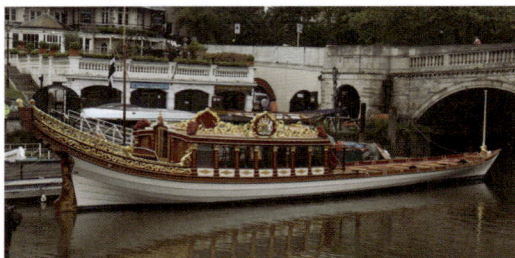

Rod's role at the Queen's Diamond Jubilee on the River Thames included surveying the Royal Barge, Gloriana while a fleet of ex-RNLI boats also took part

Captain Shaw is recognised as having extensive knowledge and experience in maritime safety. As a consultant adviser to many governments and international ports and harbours, his expertise has been sought from the Falkland Islands to Russia, and his participation has resulted in enhanced safety in ports and coastal areas.

Captain Shaw was a member of the Harwich Lifeboat Station for over 35 years, and in 1990 Rod Willis Shaw published a book *Launched on Service*, to celebrate the revived station's 25-year history on the centenary of the first steam powered lifeboat.

Rod was appointed MBE in the 2013 Queen's Birthday Honours for his contribution to maritime safety. He said at the time:

> "I'm really proud to have been recognised in this way by Her Majesty, particularly as it carries on a family tradition, as my mother was also the recipient of an MBE (for services to local education). I'd like to accept this award on behalf of all my colleagues in the MCA who work tirelessly to improve maritime safety and who shared the responsibility for inspecting over 500 vessels that were participating in the Diamond Jubilee Pageant."

The moment of truth — Captain's log

"Once delivered to Harwich, the *W&S* was craned onto blocks on the quay at Gas House Creek, then occupied by A.J. Woods engineering company. Its early days in the yard at Gas House Quay were productive. Tony Woods and his staff were most cooperative and provided good facilities for the first phase of stripping the old paint, re-caulking the

hull and dressing the extensive steel keel bar. Before being caulked and anitfouled, the hull was scraped back, revealing the beauty and craftsmanship of its double-diagonal construction, which remained sound after 85 years.

Coincidently, over 50 years earlier I had picked up some useful skills in Gas House Creek when I assisted shipwrights from Cann's Shipyard. I was the 'Dolly Boy', required to crawl beneath the wooden boats under construction and repair, and hold the Dolly so that the wooden planks could be clinched up with special copper nails. During my long career rising through the ranks to become a Master Mariner I learnt many other skills for wooden boatbuilding and repair.

During inclement weather a start was made in the engine bay and here we faced the first minor setback, although it was predictable. The six fuel tanks contained a considerable amount of diesel fuel, most of which was found to be contaminated, and in some tanks the wax was so thick it would not allow the contaminated fuel to be drained out. This was eventually resolved by cutting open the tanks and allowing the residue to drain into the boat's bilges which was then scooped into portable containers for safe disposal ashore. This alone took a number of months, more than had been envisaged. After this the engine bay had to be completely de-greased and decontaminated.

2013 SPRING   The actual condition of the two six-cylinder Ford D-series engines was unknown, other than they had been flooded with rain water when the boat lay ashore. Specialist advice was sought from a professional diesel engineer with experience of the Ford Dorset engines. Peter Coleman Engineering advised that the engines should be removed from the boat so that they could be stripped down and fully restored with new parts. Peter removed them to his workshop and has totally restored each engine so that they are now ready to be reinstalled.

The condition of the engines was unknown, but at first sight they didn't look too healthy. And the fuel tanks (above) definitely needed replacing

A major setback was Tendring District Council taking back possession of the Gas House Quay and, following a couple of months grace, putting pressure on us to remove the boat from the site irrespective of its condition. Any attempt to remove the boat, which was in a very vulnerable condition, could have had a disastrous effect on the project. With the help of the Harwich Lifeboat crew, especially Cox'n Paul Smith, Mechanic Davy Thomson, crewmen Darren Priestnal and Paul Griffin, the boat was hurriedly prepared so that it could go into a mud berth.

Another crane was called for to drop ON736 into the muddy creek. The water ingress was excessive when it was first lowered into the water but that was soon reduced to a trickle as the double mahogany hull planking tightened up. There were two petrol-powered bilge pumps

The body of the Boat is the mahogany hull
which, when exposed after 90 years,
showed the quality of material, its durability
and the level of craftsmanship in its build

working flat out for several weeks.

2014 AUGUST   Meanwhile, extensive work continued in the engine bay. All traces of the old contaminated diesel were removed and the bilges cleaned to ensure there was no environmental pollution when the bilge water is pumped out. In the early days you had to ditch your boiler suit after each session as your clothes were beyond washing.

    It was decided to reinstate the fire resistant bulkheads fore and aft of the engine room. These are now well advanced and special fire retardant paint is being applied to the sides of this compartment, which is about 12ft square. This should have been applied in time for the 2014 Spring weather, when caulking the main deck was to be the first priority. The project, which was envisaged to take two years, was already behind schedule due to discussions about vacating the council-owned site. It was still within budget, however.

    During 2014, the engines that had been sent away for professional reconditioning, were now back and being connected up to the shafts. The decks had now been caulked so rain water was no longer a problem.

2015 NOVEMBER   Two years from the start of works marked a significant milestone for us. The engine room having been flooded for so long, the contaminated diesel, the six fuel tanks — many severely corroded — the batteries which had imploded (or is it exploded ?) means we have spent probably 12 months completely dedicated to this area of the boat.

    The hull is finally watertight so the mud berth has done its job.

Because of the engine room work I suppose we must be six months behind envisaged schedule. Had we been in this position at Easter we would have had all summer to do the wheelhouse and upperworks. So long as we keep going forward and within budget is all we can ask for.

In 2015 the W&S was awarded national recognition and placed on the UK Register of Historic Vessels, Certificate Number 2996.

Completed extensive engine trials on February 13, this week's Spring tides. Have a bit of a problem on the port gearbox but believe this can be resolved. Hull is now pretty sound and she could lay afloat all the time if necessary. Starting to replace the hydraulic steering, the ram was completely rebuilt by the shipyard in Hull (Humber) and need to reposition the fluid reservoir which sat in the middle of the main console, new hydraulic hoses obtained for rope locker.

Other jobs in hand:
Continued replacement of the stanchion mountings to deck on port side
Reinstatement of kick board from port bow to midships.
Starting to reposition the new starboard side fuel tank.
Onboard discussion with a woodworker who has a lathe to assist shaping

.The sorry state of the deckhouse is clear to see from this angle

heavy timber at starboard bow fairlead.

Starting to cut & prepare new deck board in main cabin. There will need to be a hatch to access starboard sea valve underneath.

Painted trunking in Master's cabin in way of hydraulic hoses being re routed.

Started preparing main genset in the hope the Cummins engineer will be in the area soon.

Took fuel for main engines and small genset. Disposed of waste etc, etc.

Since finding the ex-Penlee Lifeboat the renovation pathway to bring it back to life has been chequered but has mainly been a case of two steps forward and an occasional step backwards.

2016 OCTOBER At the moment doing a lot of odd jobs to prepare for the week commencing Monday 14 November when we shall be conducting extensive river trials. During that week we shall be doing steering trials to test

Rod (left) with Davey Thompson and Paul Smith, preparing to lower the boat into the mud during the early stages of renovation (right)

the replacement hydraulic system. Will also be doing prolonged engine trials with both engines working at full power. And hoping to get the onboard genset (ex Thames Clippers) working so that we can commence rewiring and installation of the electronics.

This week have started the refurbishment of the master cabin. I need to replace the roof as this has rotted through.

Works during the winter

Decided on complete replacement of the after cabin which had

208

deteriorated beyond repair.

Rear seat incorporated on back end.

New hatch included not only to provide escape hatch from Master Cabin but also to provide natural light in this large cabin.

Removed mast bracket which was found to be quite lightweight when you consider the aft mast with radar scanner she used to carry.

Installed solar panel system.

Some deck caulking completed on back end where new cabin joins deck.

Installed 3 permanent bilge pumps in the 3 main compartments.

Replaced portable genset.

Some temporary electrics completed: 12v to bilge pumps & 240v cabin lighting to 3 main compartments.

Commercial genset loaded before wheelhouse replaced, repositioned in engine room.

Both engines run up on each Spring tide. Extensive work to attempt resolving port gearbox problem.

Additional lighting in engine room installed.

Engine room shelving provided to assist spares stowage and engine tools.

Preparation works on renewal of starboard bulwark.

Timber preparation for kick board replacement.

River trials to test engines under load and extensive steering **2016/7 WINTER** manoeuvres.

General maintenance works of restoration previously completed.

Restoration progresses. As we still had port gearbox problems, I decided to resolve it once and for all, so took the gearbox out, which is no easy job, and took it to Coventry to a gearbox company that knew all about these old mechanical boxes. In fact they had the old records of these boxes which were made by Hawker & Sidley, they were able to tell me when my gearbox was sold and who to. They took the box completely apart and found many problems, mainly corrosion from when the box/ engines were flooded in the field with rain water. I had paid a local company to overhaul the engines and gearboxes when we moved the boat to Harwich from Falmouth; clearly they did not do a good job and, in fact, it's questionable that they even took this gearbox apart.

2017 MAY   Going back to Coventry on Tuesday to pick it up and then with Paul Smith and with help of the lifeboat crew will endeavour to get it back in, and then hopefully we will have two engines fully operational and can get back to all other works. Have just come back from a two day training course on small boat plumbing and electrics at Lowestoft Boat Building College. Was very good, learnt a lot. You are never too old to learn!!.

2017 JULY   Had got the reconditioned gearbox back from Coventry and eventually fitted, so this week we wanted to do running trials and also get lifted out at Shotley Marina to check the hull and antifoul. We had other important jobs such as changing a sea valve and a number of engineering jobs. All in all a busy programme which we have just completed. Was short of crew as no one was available to help, so my brother in law and I had to do it all which made it interesting when trying to manoeuvre and tie up at the same time. Bearing in mind I have not got the Morse cables connected yet so we are still operating from the engine bay. I do miss Davey Thomson to assist with this: the ex-station mechanic has returned to his homeland as Mechanic on the Tynemouth lifeboat. Anyway, all done.

2017 AUGUST   Now that both engines and gearboxes are up and running I will be able to get back to the topside restoration. Another problem that has occurred is that the plywood that I have fitted and treated very well with primers and undercoats has deteriorated quite quickly. To be fair, the timber yard had warned me that it all comes from China now and is of poor quality. So I need to go back to the wheelhouse roof and cabin top as they have started leaking. In order to keep some 1930s standards, I was planning not to use modern plastics but now, because of the poor quality plywood, think I will have to use modern materials.

We are still in Gas House Creek, and I'm not sure what's happening on the other side of the creek. Harwich International Port (the landowners) keep saying they are going to set up a fence. But there is still no activity on my side so I just carry on day by day. The boat is now tight, the hull has tightened up wonderfully with no ingress of water so if

I can stop the rain water coming in I will have a dry boat. Still have a lot of work so am under no illusion, although the running trials this week have given me more reassurance that I am making progress.

Have had the boat out at Shotley for the last two weeks. Been working on the fenders plus a repair to the stem post which could not be done in the Creek. Also took the opportunity to antifoul, plus painting the topsides.

During inclement weather have been working down below on the engines, which still take a lot of time even for regular maintenance and odd jobs. Looks like I may have a head gasket gone, which will be another setback. Awaiting a marine engineer to come and give me his opinion.

Took a photo (below) which shows her nice and tidy, probably its best condition since it came out of the field.

Another year, another docking

Planned to take advantage of the two Spring tides in April. Arranged with **2018 MAY** Shotley Marina for lift out on weekend of 15/4 to be told they do not lift out/in at weekends. So, planned to go across on the Monday with lift out on Tuesday 17th. Tuesday was quite windy so the transfer from lay-by berth to under the hoist was 'hairy'. Ended up putting a hole in the starboard quarter as the wind blew us back onto a corner of a pontoon which was not fendered.

On lift out the boat looked 'rough' and there were a number of comments from the hoist operator about 'wreck' or other similar comment. They positioned the boat and chocked it right at the entrance to the marina from the road, so every visitor came face to face with the boat — fortunately on its port side so the damaged hull was not readily apparent.

Shotley jet blasted on the Tuesday so we left it till Thursday before applying anti foul (thanks to Tony & Alan). We used a roller brush and it went on well. During the next 10 days worked hard every day going to/fro on the ferry. Tony & I did the blue topside which did not take too long as the surface area is not too great.

During this 10 days I planned to replace/repair the rubbing band, this took longer than anticipated, I also had the damage to repair and this needed some special attention to get a watertight repair. The plan

had been to go back in the water on Friday 27th as I had Julian booked to be on board for the extended running trials on the Monday and Tuesday. The long-range forecast was not good and we would not have a berth on Harwich Quay on the Monday due to the very strong (Force 7) Northerly gales. Discussed it all with Shotley who said they would not launch in the predicted strong winds. Decided to leave the boat on the hard standing which meant we lost the Spring tides so it would be another 10 days at Shotley. Julian did all his engineering works on the hard standing, it just meant we could not do extended running trials which was a big disappointment. The wind was as predicted and was particularly strong from the North. No harbour ferry running and excessive swell on the Harwich side.

In the next 10 days was able to continue with the rubbing band and got most of it completed. Was able to do other works such as the bow stem post repair which included cutting out a damaged section and scarfing in a hard wood block, shaping to fit and then replacing the metal stem post which needed new screw fittings. This extended docking allowed all jobs to get done including the new (2inch) white line which clearly sets the hull off.

2018 MAY During the stay on shore there were many visitors wanting to know what the boat was and its history. Also, the marina staff were impressed with the way it had come up with the hull being painted. Went back in the water on Monday May 14, the wind was quite strong from the NE so agreed with Doogan to have an inside berth on Harwich Quay.

Had this fear about a head gasket gone on the starboard engine. Was not intending to do any running trials until this possibility resolved. Alan Sharp came onboard to look and was of the opinion it was ok. The oil coming from the gasket was normal for these engines. Based on this I decided to run the engines for extended period, albeit alongside.

Had about 1.5 hours underway from Shotley as we went up river to scatter ashes. When on the berth at Harwich Quay ran both engines so a further 4.5 hours on fast idle, this meant both engines had run for 5 hours, all was well, normal temps, good overside water cooling etc.

Next day, Tuesday 15th did some running trials in the harbour for about 1.5 hours before going in the creek. Ran well in the harbour, steering still not perfect but running better since Julian did all the grease nipples and bearings which were completely dry.

Tied up on the berth having been away for over a month.

Julian's job sheet
Start Cummins genset:
>Turned over on battery, no fuel connected so did not start but turned over easily.

Oil leak on stbd gearbox to sort:
>Hose connector stripped down and hydraulic washer fitted.

Monitor when underway.
Adjust idle speed on stbd engine:
>Attempted but with no success. Thinking the engine was to be stripped for the head gasket we left it but later decided to

leave the engine so this job still outstanding.
Stbd engine manifold to tighten:
> Julian suspected head gasket gone. Alan Sharp inspected & considered ok. No further action.

Check Jabsco pump impellor:
> Discussed and decided to leave as previously inspected.

Inspect rudder glands:
> A number of grease nipples replaced as they appeared seized. Glands also appeared dry of grease.

Grease all steering nipples:
> All nipples and glands given good grease. Steering did appear easier on trials.

Inspect fuel filter:
> Port fuel filter stripped down for inspection. All appeared clean and in good condition. Replaced. Decided not to strip down the stbd filter.

Paint Spec.

Decided to keep to same paint manufacturer (TEAMAC) as previous paint was holding up well. Ordered Anti Foul and Oxford Blue from ProMain.
One coat of antifoul all round took about 3.5/4 lts.
The Oxford Blue took about 2.5 Lts.
> Both applied by roller and went on well.

Used Alan Perks and Tony for painting. On rudder and other metal work used Hammerite "Wild Thyme" (smooth) and looks good.
The white line around the hull is now a 2" line and takes a bit of doing
> as the masking tape is not 100% successful.

Fuel Consumption

As this was the longest the engines had been run continuously (6 hours), was able to get a fair idea of fuel consumption.
During the 4.5 hours running at fast idle alongside the fuel consumption for both engines was 1 gall/hr. This equates well with previous information (Scum's boat which has same engine) of 1.6 galls for one engine underway.
To summarize :
> 2 engines on fast idle tied up alongside (no load)=1 gall/hr

> 2 engines underway on full ahead (full load) = 3.2 galls/hr

Therefore. Harwich/Falmouth - 367' @ 9kts = abt 41 hours
> steaming = 131galls which equals about 600 lts.

The 2 tanks onboard hold 470lts each, say 450 = 900 lts in total.
Allowing for low level and allowing say 10% reserve, a working capacity would be about 800 lts so adequate to do Harwich/Falmouth with some reserve. Therefore both tanks need to be installed and connected.

Costs

Shotley Marina has varying costs. Springtime (April/May) when everyone is getting their boats in the water is their most expensive time. Always enquire when their reduced costs are (possible August).
The lift out/in is £280 each way + £80 pressure wash + £70 for blocking up. ALL PLUS VAT !
Believe these costs include some yard time but when weather delayed and had another 2 weeks in the yard they charged £100/week for hard

standing. Total costs to Shotley Marina which included 4 weeks ashore about £1100 !!!

2018 JULY The picture of the *Michael Stephens* looks good and makes me wonder if I should not have returned ON736 back to original design. Anyway its too advanced now, am presently recovering the wheelhouse and accommodation in the dark oak style which gives it some character compared with the all white. 'Griff' is currently working on the bow pudding fender which it originally had. Also had a chap volunteer to do sign writing so he is going to do the RNLI flag on each bow, which I see *Michael Stephens* has and it does set the hull off.

I recently purchased a new battery which was different to what Davy supplied like they have on the Severn. However, these Severn batteries are part of a bank so they are hooked up in parallel, which means each individual battery is less 'beefy'. When I talked to a specialist battery firm up North they recommended a specific heavy-duty battery which is about half as big again but is really doing the job so feel I have satisfactorily concluded this important function. Will now order a battery for the port engine but I have to go to Shotley or Halfpenny Pier to load it up.

I also had some starting problems which I had put down to low battery power but I eventually concluded it was the starter solenoid playing up. I have now replaced this and although its early days it appears to have resolved my starting problem. Will be flashing up on each Spring tide this week so will know more by the end of the week.

Its Sea Festival today but I am not involved by choice. Perhaps by next year will have ON736 ready for public display but I'm not trying to make deadlines. I will work as and when it suits me, which is most days, the project has been a retirement saviour and I am so pleased I took it on. It would be nice to get it back to Penlee but I'm not making promises as yet.

Have had the boat round on Halfpenny Pier in order to fit the bow fender which young Griff had reworked because it was light blue in colour which did not look right. As you can see in the photo he has modified it and it looks much better.

In the good weather this summer I have recovered the back cabin. The ply was delaminating so I have had to recover and seal it with plastic sheeting so that it does not absorb the moisture. Could have done without this task because there is still so

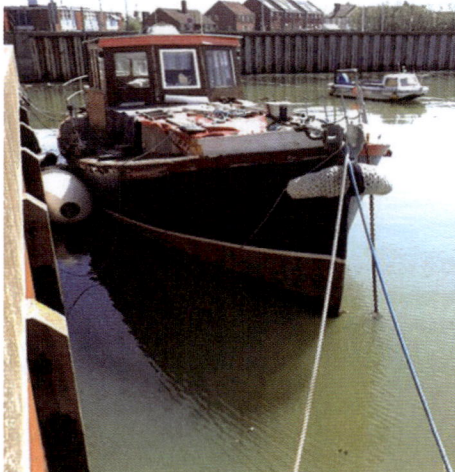

much to do, but hopefully this is now complete and the back cabin is fully weathertight.

Touch wood, the engines are all performing well, so that is one **2018 OCTOBER** achievement and am preparing to do the rewiring over the winter. Have appointed a 12v motor specialist to advise me on the necessary circuits. This has all come about because when we took the engines out the engineer just cut every cable. Not only that, but the old switchboard and fuse panel were in dire need of replacement so it won't do it any harm to install a complete new circuit.

This week I have started on the decks, should have done it earlier in the summer. They were completely dried out, to such an extent the rain water just seeped through. I have started a course of raw linseed oil to try and stabilise the oak planking.

Chris Britton, who used to be on the Harwich lb crew, is Chief Engineer on Svitzer tugs and his family have always had boats as Harwich fisherman. He came onboard and has offered to help, which is good because I lack engineering knowledge. As an example, he noticed I was using the wrong grease in the shaft. That's great, just what I need, so hopefully he will become a useful adviser.

Anyway, I continue as and when weather permits. Have just completed the after peak which was a big job as it needed complete restoration, not having been touched since its RNLI days. Attach a photo of the fore peak, but the after peak was much more detailed as the paint work had to be taken back to the wood, the old floor was rotten and had to be replaced. I have added racking in order to stow the tow ropes etc.

I'm about to replace the starboard bow bowser which I have rebuilt in full. Will get a photo of this. Otherwise I'm having a good clear up before starting the electrics, which

I have been planning to do on more than one occasion previously and got delayed on other works. Would like to think the engines are now complete. I'm just completing some new deck boards in the inside cabin so plenty still going on.

2020 JANUARY   I understand Gordon Burns is upset by my referring to the boat as 'abandoned'. I look through some of the early photos and you can make your own judgement about the condition of *W&S* when I got her. The fuel tanks alone tell a story of neglect and the picture of her in the yard is clearly abandonment. Look at the picture of the engine, would you go to sea in this condition? Sorry Gordon, 'abandoned' is my summary. I don't believe she would have survived another winter in the yard without extensive restoration, which would have made it financially unacceptable, so she would have ended her days at Falmouth.

2020 FEBRUARY   I have been clearing out the midship bilges recently because I was laying new floor boards, so I decided to do this whilst the bilges were open. Found a very corroded spanner which is probably original. Other than that I am having a damn good clear up and getting rid of all the rubbish. It all takes time. Had the yard still been open I would have made much more progress. Its like the rubbish, it takes up considerable time when I could be doing the rewiring, as just one job that remains on the list. Completed the after peak rope store, very happy as it gives me much more storage. Also installed a new hatch, which meant cutting some deck away.

2020 APRIL   Good photo from Quinton of the boat underway [on page 193]. This is similar to what I am trying to achieve. Same hull colours. Probably not the mast, although I have retained the mast step bracket should I consider a small mast. Wheelhouse is more varnished but may reconsider this as it is not standing up too well in this weather.

Both engines run on the Spring tides last weekend, all good in that department. Still in early stages of rewiring. Decided to rip all old wiring out and start again as there were many connections which have deteriorated when the vessel was abandoned — Sorry, 'laid up'!

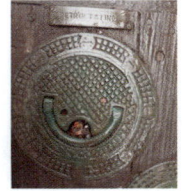

Renewing all the control cables onboard, just doing the engine stops which for some reason have never been connected. Having to keep a close eye on the weather as it can blow straight down the creek making it very lumpy on the berth.

2020 AUGUST

Since the 'lockdown' due to the coronavirus pandemic, I have brought many items off the boat for servicing at home; stanchions, turn buckles, shackles etc. At least I am making progress, although with this good weather I could be doing much more on the boat, if only getting it ready for the Shotley Marina lift out. I have moored the boat port side to, while I replace the starboard rubbing band. Have had to source decent oak to replace the original belting which has deteriorated beyond acceptability. Am also showing photos of the official number and the original petrol filling port, one of the few original fittings remaining.

Modern youth shows no respect for history

2020 SEPTEMBER

Another chapter for the book. On Tuesday, **September 22**, *W&S* was vandalised! They did a thorough job and smashed everything in sight. It's getting wide publicity; I've had SW News Agency on the phone. Many of the original fittings have been damaged, even the port lights set in the deck were smashed, Some photos show what Harwich yobs can do. Damage estimated at £5000+ and probably put me back 18 months at least. Every bit of glass was smashed. Demoralised!

Have had good response from the lifeboat enthusiasts. At least they were all sympathetic! It would be nice to have some Mousehole

connection to the restoration, even if it's only a lifebuoy or compass with a Mousehole inscription.

2020 OCTOBER

With the insurers sending an assessor they must be expecting to make a substantial payment. We have no local boatyard, so I have to consider the best option for progressing the works. The kids emptied the fuel tanks into the bilge so that is going to take some cleaning and shore disposal of the bilge water. I will need to see what happens with the insurance claim and plan from there.

The insurance assessor has been to inspect and was very helpful.

2020 NOVEMBER Had good contacts on where I could get individual windows renewed. He deals with canal boats and the like so was understanding of my situation. He will now report back to the insurance company, who will decide how much to pay out. Most of the work: clearing up, engine bilges to clean, all glass (27 panes) needing to be replaced is not really shipwright work that would warrant me taking the boat to Woodbridge or Maldon. All the works are within my capability, it's just the time it would take me.

Thanks to a local resident who saw the group of three or four boys fleeing the scene, and took a photo, the police know who the main culprit is: a lad of about 12 years old. However, he cannot be interviewed because his mother is not 'a responsible adult', whatever that means. Police did not say how they would progress this. Damage is now estimated to be £10k. Fortunately I have receipts for the wood thrown overboard but how do you account for tools which have sunk in the mud?

2020 DECEMBER I was a little reluctant to start replacing the glass because the kids are still around, what happens if they come back to smash the new glass? I may leave it till after Xmas. Tried to start the engines up yesterday and found the port engine Morse control jammed. I know they had been playing with the handles so I must now report that to the insurance surveyor.

Still hoping the insurance will cover the expense, rather than begging with a 'give-me' fund or whatever its called.

Police now say they have interviewed the suspect and will be talking to him again, and let me know what progress they make.

Have not heard from the insurance surveyor for some time now. What a way to approach her 90th birthday!

Back on course

2021 FEBRUARY Eventually, one week before the boat's 90th birthday, the insurers settled on a figure that covered most of the materials. Have managed to source replacement windows but that still leaves months of work to return the boat to its previous condition. Now reckon the episode has set me back about one year. Nevertheless, the *W&S* still has a lot of life in her and there is a lot of love being offered in support.

2021 MAY One of the most frustrating aspects of this vandalism escapade has been seeking suppliers who could produce custom parts to replace the 1930's fittings. Of the many enquiries I made three companies have been most supportive and used their expertise to replace like-for-like fittings, particularly the glass ports which had to be engineered to within 1mm. Harwich Glass, Halmshaw of Beverley/Hull and Eagle Boat Windows all need to be commended as without them ON736 (*W&S*) could not have been restored to pre-vandalism standard.

The boat in better days before the vandals struck. And the mobile workshop installed on the quayside

The downtime pending the insurance settlement has necessitated a project management rethink. Although it involves additional expenditure it has been considered necessary and practically prudent to have a mobile workshop so that all the glass replacements and specialised tools such as wood router could be stored alongside ON736's mud berth. Cleaning the contaminated bilges and getting rid of all the shattered glass also needs coordination of resources.

Windows in the wheelhouse replaced, fore and aft, and ready for more river trials

This is obviously an ongoing, long-term project and despite inevitable ups and downs, I am determined to see *W&S* (ON736) fully functional and in a serviceable condition for whatever role awaits her. In the context of nine decades a few more years is a mere blip. As we celebrate the boat's 90th anniversary, the resurrection continues and this Nautobiography approaches its publication date.

I thank everybody who has expressed interest and support for the project and hope to welcome you aboard when circumstances permit. **»**

———————— **Rod Shaw, Harwich, 2021** ————————

The project to bring 'our' boat back to life has been of national interest. The Chairman of the RNLI Sir Charles Hunter–Pease took a specific interest and visited the project to get personally updated on the progress of the restoration. He commented that this famous lifeboat (ON 736) was worthy of saving as it had a place in RNLI history due to the number of lives saved and the outstanding rescues the boat had completed.

Acknowledgements

In addition to those mentioned in the Foreword, special thanks go to many people in Penlee, especially Neil Brockman, who served on the *Solomon Brown* (which went down with his brother Nigel in 1981) and was Cox'n from 1992-2008, receiving a Bronze Medal in 1994. Neil's successor, Cox'n Patch Harvey was most welcoming to the author on visits to Newlyn. Particularly helpful during the book's production was Amy Smith, lifeboat crew member and photographer, who helped prepare many of the images for reproduction. The original cover design was fine tuned by Konrad Manning. Megan Green helped to keep the project (and the author) on an even keel and kept the pages turning during the Covid pandemic, and thanks to the Booths Print team for their patience.

I made good friends with the crew of Harwich lifeboat *Albert Brown* (17-03), Cox'n at the time Paul Smith; mechanics Brian Hill, Andrew Moors and Davey Thompson, also the Press Officer Keith Churchman, a proud Cornishman and ex-crew member of the St Ives lifeboat, Tony Woods of A.J. Woods and Tony O'Neil of Lightship LV18.

In Falmouth I have been made more than welcome by the maritime community, especially the ex-pilot Dave Pickston, salvage expert Brendan Rowe, who took me round the Lizard and across Mount's Bay, lifeboat Cox'n and Falmouth Dock master Luke Wills, Cox'n Andy Jenkin, who gave me the helm of *Richard Cox Scott* (17-29) during an exercise with my own boat, Diccon Rogers of KML, Penny Philips of A&P, tug skipper Dave Hughes, boatman John Pill and family, Jim Lloyd-Davies. And we owe a special thank you to Iain MacQuarrie of MacSalvors for moral and financial support.

Picture credits

Original photography: Graeme Ewens Cover and pp 33,35,40,50,82,199(br),200(tl),202,204-209, 214(top). Rod Shaw 194(br)195 (b),210,212,215-217. CISMA 49,115. Megan Green 197, 203. David Hughes 9, 170,171. Chris Yacoubian 54, 160 (b), 177. 'Flash' Harry Welby back cover (t),62-3, 104. Shirley Bodèll 187-189, 192. Quinton Nelson 193, 194, 195 (t)

Archive photography: RNLI, Penlee lifeboat, Morrab Library, Penlee House Museum, Museum of Cornish life photographs.html, Kresen Kernow, Iain MacQuarrie of MacSalvors. Images from Scotland were provided by the relevant stations and Cromarty Archive. Many historic images came from personal collections of crew members and their families, especially Edwina Reynolds, granddaughter of Eddie Madron, the Blewett family and Nim Bawden. Several images of casualty vessels were found at wrecksite.eu, on open access websites, or were forwarded by friends and colleagues, and these have been credited where possible.

Online and digital sources quoted

The primary resource for the Penlee lifeboat is found at www.rnlipenleelifeboat.org.uk

The *Lifeboat* Journal (Magazine) lists shouts from all stations at www.lifeboatmagazinearchive.rnli.org Quinton Nelson can be found at www.nelsonboats.co.uk.

Mark Waltham has an informative site at www.lifeboat-laura-moncur.co.uk

Information on almost any wrecked or sunken vessel can be found at www.wrecksite.eu

Richard Larn, co-author of the *Shipwreck Index* and many other books on shipwrecks hosts www.shipwrecks.uk.com. With his wife Bridget he started the Charlestown Shipwreck Heritage Centre at www.shipwreckcharlestown.co.uk

A good source of information on warships is https://uboat.net and another useful site is www.warssailors.com/freefleet/index.html

www.shipsnostalgia.com has many threads including detail about HMS *Warspite*.

More on *Warspite* can be found at https://freepages.rootsweb.com/~treevecwll/family/viditorshw.htm

The Cornwall and Isles of Scilly Maritime Archaeology Society is at www.cismas.org.uk

Page 98, www.ournewhaven.org.uk has the life story of Capt Frank Gilbert (Taycraig)

Page 112-113. *S.O.S.* This film can be viewed at http://film.britishcouncil.org/sos.

Page 124. The story of the convoy attack is found at https://johnknifton.com/tag/m-v-polperro. The basic research having been made freely available by David Betts. http://www.cornishman.co.uk/Grandson-writes-fateful-

Page 128-9 The Day A U-Boat Hit The Wolf is an edited version of a longer piece by Lynda White, nee Cherrett, granddaughter of Charlie Cherrett and recorded at: U-boat Archive

Page 147. The full version of Ralph Richards' statement is at www.porthlevengigclub.com/history

The full story of the St Ives lifeboat disaster can be found at www.facebook.com/Dive.St.Ives/posts/ 315579912419856 And also at www.facebook.com/781224208/posts/10156378848149209/ #IronMenInWoodenBoats #GoneButNeverForgotten

More general background and foreground detail can be found at Nostalgic Penzance & Newlyn, a Merchant Navy Message Board and at www.thisisplymouth.com

Neville Green, One Cornishman's Life, http://www.lizard-lives.uk

http://www.mariners-l.co.ukhttps://uboat.net/allies/merchants/ship/957.

https://transportsofdelight.smugmug.com/SHIPS/British-Coastal-and-Short-Sea

https://discovery.national archives.gov.uk/details harwichanddovercourt.com

Newspapers and periodicals:

The British Newspaper Archive
The Lifeboat Journal
Lloyd's List
The Cornishman
Western Morning News

Cornish Evening Tidings
The West Briton
Penzance Gazette
Dundee Courier
Liverpool Echo
The London Gazette January 29, 1858

Bibliography

Barnicoat, David and Culliford, Simon
  Falmouth Lifeboat, 150 years of saving lives at sea. Falmouth RNLI, 2017
Bathurst, Bella
  The Wreckers, Harper Collins, 2005
Beherna, John,
  Westcountry Shipwrecks, David & Charles, 1974
Bird, Sheila
  Mayday! Preserving Life from Shipwreck off Cornwall, Ex Libris, 1991
Bristow, Alan (with Malone, Patrick),
  Helicopter Pilot, The Autobiography, Pen and Sword Aviation, 2009
Browns Nautical Almanac 1938
Camidge, Kevin,
  St Anthony, A Desk-based Assessment, CISMAS, Penzance, 2013
Camidge, Kevin & Randall, Luke
  An Archaeological Survey of Mount's Bay, CISMAS, 2009
Campey, Rachel.
  Penlee Lifeboat Station. RNLI, 2017
Carter, Clive.
  Cornish Shipwrecks: The South Coast, Vol 1, David & Charles
  The Port of Penzance. Lydney: Black Dwarf, 1998
Corin, John and Farr, Grahame.
  RNLI Penlee. Penlee and Penzance Branch of RNLI, 1974/1983
Course, Capt A.G.
  The Deep Sea Tramp, Hollis & Carter, 1960
Fowles, John and The Gibsons of Scilly.
  Shipwreck, Jonathan Cape, 1974
Fry, Eric.
  Life-Boat Design and Development, David & Charles, 1975
Jeffrey, Andrew
  Standing Into Danger, Dundee Branch, RNLI, 1996
Knight, Gavin.
  The Swordfish and The Star. Chatto & Windus, 2016
Larn, Richard and Bridget,
  Shipwreck Index of the British Isles Vol 1. Lloyd's Register, 1995
  Shipwrecks Around Mount's Bay. Tor Mark Press, 1991
Larn, Richard and Carter, Clive,
  Cornish Shipwrecks, The South Coast, David & Charles, 1969
Leach, Nicholas
  Harwich Lifeboats, An Illustrated History, Amberley, 2011
Marshall, John
  Royal Naval Biography
McCormick, W.H., *The Modern Book of Lighthouses, Lightships and Lifeboats,* A&C Black, 1936
Noall, Cyril,
  Cornish Lights and Shipwrecks, D Bradford Barton 1968
  The Story of Cornwall's Lifeboats, Tor Mark Press, 1970
Noall, Cyril and Farr, Grahame

Wreck and Rescue Round the Cornish Coast II, The Land's End Lifeboats, III, South Coast Lifeboats, D Bradford Barton, 1965
O'Byrne, William Richard
  A Naval Biographical Dictionary
Pearce, Cathryn J.
  Cornish Wrecking, 1700-1860: Reality and Popular Myth, Boydell Press, 2010
Perks, Richard-Hugh
  Sprits'l. A portrait of sailing barges and sailormen. Conway Maritime Press, 1975
RNLI
  Regulations of The Royal National Lifeboat Institution, 1950, amended 1957
Sagar-Fenton, Michael.
  Penlee, The Loss of a Lifeboat. Bossiney Books, 1991
  Penlee Lifeboat - The First 200 Years. Penlee Branch of RNLI, 2005
Sawyer, l.A. and Mitchell, W.H.
  The Liberty Ships, David & Charles, 1970
Shaw, Frank H.
  Famous Shipwrecks, 1930
Shaw, Willis (Rod)
  Launched on Service, Harwich Lifeboat Station, A Quick & Co, Essex, 1990
Smith, C Fox,
  Adventures and Perils, Michael Joseph, 1936
Smyth, Admiral W.H.
  The Sailor's Word-Book, Blackie 1867, Algrove 2004
Tangye, Nigel
  From Rock and Tempest, William Kimber, 1977
Tregenza, Leo.
  A Harbour Village, Yesterday in Cornwall, William Kimber, 1977
Trengrouse, Henry,
  Shipwreck Investigated. . .and a remedy provided, James Trathan, Falmouth, 1817, reprint 2018
Trinity House
  Visiting Committee Inspections, West Coast Station, nd
Vivian, John.
  Tales of the Cornish Wreckers. Tor Mark Press, nd
Wakeham, Geoff,
  Royal Naval Air Station Culdrose 1947-1997, John K Miln nd
Warner, Oliver,
  The Life-Boat Service, Cassell, London, 1974
Whymper F.
  The Sea: Its Stirring Story of Adventure, Peril, & Heroism. Vol l & ll, Cassell, Petter and Galpin, c1885
Williams, David L.
  Maritime Heritage: White's of Cowes. Silver Link, 1993
Williams J.A. & Gray, J.B.
  HM Rescue Tugs in World War II. HMRT Veterans' Association, nd
Wilson D. G.
  Maritime History of Falmouth, Halsgrove, 2014
Woodman, Richard,
  Keepers of the Sea, Chaffcutter Books, 2005

Selective **INDEX** *of casualty vessels, featured lifeboats and significant references*